Space Security Law

Ruwantissa Abeyratne

Space Security Law

Springer

Dr. Ruwantissa Abeyratne
International Civil Aviation Organization
999 University Street
Montreal H3C 5H7, Quebec
Canada
tabeyratne@icao.int

ISBN 978-3-642-16701-0 e-ISBN 978-3-642-16702-7
DOI 10.1007/978-3-642-16702-7
Springer Heidelberg Dordrecht London New York

Cover design: WMXDesign GmbH, Heidelberg, Germany

Printed on acid-free paper

Springer is part of Springer Science+Business Media (www.springer.com)

Preface

By any modern standards of human endeavor and research, communications made possible by global navigation satellite systems and space transportation stand preeminent in the wonderment they offer. What began as exploration of outer space in the nineteen fifties and sixties is now full blown tourism in space. Added to that is the startling possibility of the existence of life in outer space which makes us not only think but wonder in amazement. Stephen Hawking – one of the world's most eminent and knowledgeable physicists – has stated that in a universe with 100 billion galaxies, each containing hundreds of millions of stars, it is unlikely that life forms are present only on Earth. Hawking has also said:

> To my mathematical brain, the numbers alone make thinking about aliens perfectly rational... the real challenge is working out what aliens might actually be like[1] ... I imagine they might exist in massive ships, having used up all the resources from their home planet. Such advanced aliens would perhaps become nomads, looking to conquer and colonize whatever planets they can reach.[2]

Against this bewildering backdrop, we continue to use and explore outer space, take pictures, calculate trajectories of planets and determine who owns the moon and what the purpose of outer space exploration is. An added dimension is the use of aerospace in terrestrial transportation where an aerospace plane will take off as an aircraft, go into orbit, enter the atmosphere using the Earth's orbit into its destination, cutting the travel time significantly. It is said that by using this method, air travel time can be reduced drastically. For instance, a journey by air between Los Angeles and Sydney, which would now take 14 to 16 hours by conventional air travel, could take 2 hours or less. None of these technological feats would be possible without the advancement of information technology and computerized knowledge-sharing. However, with the advancement of this technology would also come the threat of cyber terrorism, which is a real cause of concern to astronomical science and space travel.

---

[1] http://www.timesonline.co.uk/tol/news/science/space/article7107207.ece#cid=OTC-RSS &attr=797084.

[2] http://www.telegraph.co.uk/science/space/7631252/Stephen-Hawking-alien-life-is-out-there-scientist-warns.html.

In March 1998, the web site of the National Aeronautics and Space Administration (NASA) of the United States received a "denial of service" attack, calculated to affect Microsoft Windows NT and Windows 95 operating systems.[3] These attacks prevented servers from answering network connections; crashed computers, causing a blue screen to appear on the computers. The attacked systems were revived, but this attack was a follow up of one in February of the same year, when, for two weeks the US Defense Department had unclassified networks penetrated, where hackers accessed personnel and payroll information.

Cyber-terrorism has the advantage of anonymity, which enables the hacker to obviate checkpoints or any physical evidence being traceable to him or her. It is a low budget form of terrorism where the only costs entailed in interfering with the computer programs of a space programme would be those pertaining to the right computer equipment.

Any interference with a space program of a nation, which would be inextricably linked to peaceful uses of outer space, would tantamount to an act of terrorism performed against international peace. The maintenance of international peace and security is an important objective of the United Nations,[4] which recognizes one of its purposes as being *inter alia*:

> To maintain international peace and security, and to that end: take effective collective measures for the prevention and removal of threats to the peace, and for the suppression of acts of aggression or other breaches of peace, and to bring about by peaceful means, and in conformity with the principles of justice and international law, adjustment or settlement of international disputes or situations which might lead to a breach of the peace.[5]

It is clear that the United Nations has recognized the application of the principles of international law as an integral part of maintaining international peace and security and avoiding situations which may lead to a breach of the peace.[6]

---

[3]http://mgrossmanlaw.com/articles/1999/. Charter of the United Nations and Statute of the International Court of Justice, Department of Public Information, United Nations, New York, DPI/511 – 40108 (3-90), 100M at 1.

[4]Charter of the United Nations and Statute of the International Court of Justice, Department of Public Information, United Nations, New York, DPI/511 – 40108 (3-90), 100M at 1.

[5]Charter of the United Nations and Statute of the International Court of Justice, Department of Public Information, United Nations, New York, DPI/511 – 40108 (3-90), 100M at 3.

[6]On 17 November 1989 the United Nations General Assembly adopted Resolution 44/23 which declared that the period 1990-1999 be designated as the United Nations Decade of International Law (the full text of Resolution 44/23 is annexed as Appendix 1 at the end of the text of this thesis). The main purposes of the decade have been identified inter alia as:

(a) The promotion of the acceptance of the principles of international law and respect therefore
(b) The promotion of the means and methods for the peaceful settlement of disputes between States including resort to the international Court of Justice with full respect therefore
(c) The full encouragement of the progressive development of international law and its codification
(d) The encouragement of the teaching, studying, dissemination and wider appreciation of international law

No treatise on space transportation should be without a discussion on the relationship between air travel and space travel in the particular context of the legal regimes and political commonalities that apply. Therefore, against the variegated background of bewilderment and cautious optimism that space transportation offers, this book begins with an exposé on international politics, the principles of which bear upon space transportation and the closeness of air space and outer space and activities that straddle both frontiers at the same time. It discusses current issues and possibilities of communications and transportation in outer space as well as the liabilities and accountability of the key players of space exploration.

Montreal, QC, Canada Ruwantissa Abeyratne
30 August 2010

Reference

Abeyratne RIR (1992) The United Nations decade of international law. Int J Polit Cult Soc 5(3):511–523

The four tasks of the Resolution have been predicated upon the fact that the purpose of the United Nations is to maintain peace and security. See Resolutions and Decisions Adopted by the General Assembly During its Forty Fourth Session, Vol. 1, 19 Sept – 29 Dec 1989, General Assembly Official Records: Forty Fourth Session, Supplement No. 49 (A/44/49), United Nations, New York, 1990, 31. For a detailed discussion on Resolution 44/23 see Abeyratne (1992), pp. 511–523.

Contents

Table of Cases

Chapter 1
The Shifting Focus

Firstly, any academic treatment of air law and policy should recognize that air law and space law are closely inter-related in some areas and that both these disciplines have to be viewed in the 21st century within the changing face of international law and politics. Both air law and space law are disciplines that are grounded on principles of public international law, which is increasingly becoming different from what it was a few decades ago. We no longer think of this area of the law as a set of fixed rules, even if such rules have always been a snapshot of the law as it stands at a given moment. Fundamentally, and at its core, international law was considered in simple terms as the law binding upon States in their relations with one another.[1] A definition of international law was first given by the Provisional International Court of Justice in 1927 in the celebrated *Lotus* case when the World Court said:

> International law governs relations between independent States. The rules of law binding upon States therefore emanate from their own free will as expressed in conventions or by usages generally accepted as expressing principles of law and established in order to regulate the relations between these co-existing independent communities or with a view to the achievement of common aims. Restrictions upon the independence of States cannot therefore be presumed.[2]

The *Lotus* case provided a basis for international law and domestic law to function as separate entities, although there could be instances where issues such as piracy *jure gentium* and others concerning diplomatic immunities could be adjudicated under a domestic law system.

The abovementioned principle was implicitly derived from the basic rule of law as it applies even today, that in the sustained evolution of humanity from troglodytes to computer wizards a central role has always been played by the idea of law the idea that in every civilized society there must be order as against chaos and anarchy which were inimical to a just and stable society. Therefore law is the glue which binds the members of a community, whether national or international, together in their adherence to recognized values and standards. In international

[1]Jennings (1990), p. 513.

[2](1927) P.C.I.J. Ser. A, No. 9, p. 18.

R. Abeyratne, *Space Security Law*,
DOI 10.1007/978-3-642-16702-7_1, © Springer-Verlag Berlin Heidelberg 2011

law,[3] the principal subjects are nation States, not individual citizens. Public international law applies to relations between States in all their numerous and complex forms, from war to satellites and governs operational policy of many international institutions. Some of the new and emergent areas of international law govern: the use of radio frequencies; communications; the availability, exploration and exploitation of resources, whether in the sea bed or in outer space; multinational corporations; trade, investment and finance; pollution, in all its forms; international crime and multinational corporations.[4]

International law and politics overlap in instances where international disputes may emerge between nations. International law has no legislature. Although the General Assembly of the United Nations exists and functions as a regulator of international policy, being composed of delegates from all member States of the United Nations, its resolutions are generally not binding on member States,[5] except in certain circumstances. The United Nations system has no system of courts except for the International Court of Justice, based in The Hague, which can only hear cases between States if both sides to a dispute agree.[6] Even if the parties to a dispute agree to come before the Court, it has no jurisdiction to make sure that its decision is enforced or followed. Thus the question has been frequently asked that, if there does not exist any identifiable institution to make law or establish rules, to explain and clarify such rules and, more importantly, to punish those who break rules, how can what is called international law be law? Traditionally, law as perceived from a purely domestic sense, is recognized as being composed of the four – Code, Cop, Court and Clink. In other words, a law to be recognized as such has to comprise a set of rules. Second, there must be a cop or policeman to ensure adherence to the law. Third, if one breaks the law, there has to be a Court which has jurisdiction to determine the conduct of the suspect and last, there has to be a clink or punishment. International law is not strictly endowed with these four Cs and therefore remains susceptible to criticism.

The considered view of jurists and judges alike, that international law is a set of rules, is embodied in the decision of the *International Tin Council Case*[7] decided in the House of Lords in 1985 where Lord Oliver observed:

> A rule of international law becomes a rule whether accepted into domestic law or not only when it is certain and is accepted generally by the body of civilized nations; and it is for those who assert the rule to demonstrate it, if necessary before the International Court of

[3]International law itself is divided into private and public international law, the former being also referred to as conflict of laws and the latter just termed International law. See Shaw (2003), p. 1.

[4]Jennings (1990), p. 521.

[5]See Article 10 and 11(1) of the United Nations Charter, which alludes to the General Assembly making recommendations to the member States. Also, Johnson (1955–1956), p. 97.

[6]See Article 36(2) of the Statute of the International Court of Justice, which calls for States Parties to the Statute to declare consensually that they recognize the jurisdiction of the Court.

[7][1989] 3. W.L.R. 969 (H.L).

> Justice. It is certainly not for a domestic tribunal in effect to legislate a rule into existence for the purpose of domestic law and on the basis of material that is wholly indeterminate.[8]

According to this decision a rule of international law has to be accepted by civilized nations to be considered as binding. The acceptance has to be demonstrated in some form or other. One of the ways of determining acceptance and adherence by States of a rule or set of rules that could be considered international law is to observe whether a rule is observed globally through a sustained period of time. The difficulty in accepting this approach is that there have been instances in recent times where such rules have been breached or not observed by States, placing the credibility of international law in a flux and the position of international lawyers in a grey area. The four main areas of international law that have been brought to question are: firstly, the basic conceptual framework of international law as a structure based on relations between States (which was seriously questioned by the aftermath of the events of 11 September 2001); secondly, the rules governing the use of force by States (which some international lawyers have questioned with regard to the United States' occupation of Iraq in 2003 and thereafter); thirdly, the legal regime of military occupation (personified by the occupation of Afghanistan and Iraq); and fourthly, the law governing the treatment of combatants and prisoners of war.[9]

These four issues in particular, which are symptomatic of events that bring to bear the need for a renewed approach to international law call upon jurists and judges to question the fundamental premise that international law is a set of rules that are adhered to by nations amongst themselves. Followers of the New Haven or Yale School of thought have maintained that law is a process rather than a set of rules.[10] Judge Roslyn Higgins has observed[11] that law is a specialized social process rather than a set of rules, which reflects a practical approach to and recognition of modern exigencies of international relations. The idea that law is a set of rules is rejected on the ground that the process of authoritative and effective decision- making does not involve the mere application of a pre-determined set of rules but is molded by social, moral and political considerations as well.[12] The realities of international relations are not reducible to a simple formula or set of principles but are dictated to by the interaction of States based on the primacy of a State and the philosophy that the world is organized on the basis of co-existence of States.[13] The interrelation of States and comity takes away from international law the common attribute which many have assigned to it, that it is a stable domain which relates in some complicated way to society or political economy or class structure. Instead, international law is now regarded as practice and argument about

---

[8][1989] 3. W.L.R. 969 (H.L) at p. 1014.

[9]Lowe (2003), pp. 859–871 at 859.

[10]Arend (1999), p. 26.

[11]Higgins (1999), p. 1.

[12]Bull (1977), p. 128.

[13]Freidman (1964), p. 213.

the relationship between something posited as law and something posited as society.[14] One commentator has even gone to the extent of recognizing that international law is merely a particular type of discourse about international social life.[15]

International law and international politics are, in a way, a type of discourse which is manifested both by oral and written communication and state practice between officials of States. This is supplemented in certain circumstances with symbolic acts of States. The discourse which occurs at international politics drives the process and development of international law, to the extent that one commentator argues, quite validly, that international discourse paves the way for the establishment of international rules.[16]

States may, through interactive discourse, either between themselves or through the United Nations or other international or regional organizations, establish international custom and practice which may mature through the effluxion of time into principles of international law. One example is the declaration by one State of its territorial boundaries. If such a declaration is not challenged and is acquiesced by other States concerned, it would represent a legal principle to be followed in the future. Another way in which a State could influence international politics through the legal process is by invoking the international institutional legal process. This process often results in pronouncements being made by the United Nations General Assembly. For example, in Resolution 788, the United Nations commended the Economic Commission of Western African States (ECOWAS) for its efforts in restoring peace, stability and security in Liberia and conversely, in Resolution 1244, the Security Council condemned NATO action in Kosovo.

The shift of focus in international law and politics is due in part to the unique nature of events of recent times, which have deviated from established public law principles of war and belligerence. States have been under a certain compulsion to interpret their own positions with regard to self defence in the face of unknown enemies and threats by groups of persons rather than States whose geographic and territorial boundaries are known. For instance, consequent upon the events of 11 September 2001, the action taken by the United States in Afghanistan was first perceived to have been against the group of persons who were deemed responsible for the attacks on the World Trade Centre and other buildings within United States territory. Therefore the military presence of the United States in Afghanistan was not against the governing authority of the State itself but against persons who had found refuge in the country. The next development was justification for the military presence against the Taliban government who were perceived as harboring persons who were likely to continue to attack the United States and her people. International lawyers and politicians are compelled to view such instances with caution and interpret them according to applicable law. For example, if the United States went

---

[14]Kennedy (1988), p. 8.

[15]Purvis (1991), p. 115.

[16]Arend (1999), p. 27.

on the basis that a sovereign State was harboring terrorists who continued to be a threat and would possibly attack the United States, its action in Afghanistan may arguably be calculated to be an act of self defense under the United Nations Charter. The invasion of Iraq in the spring of 2003 is another instance where international lawyers may argue whether the use of force was necessary, leading some to respond that such a measure was aimed at preventing Iraq from using weapons of mass destruction. This by no means implies that both actions of the United States in Afghanistan were justified under the principles of self defense as practiced at international law. However, there is conversely no cogent reason to believe that a nation under siege from terrorist attacks should wait inordinately until the diplomatic machinery took its course, particularly if the State concerned had intelligence to indicate that such a delay would be detrimental to its interests and that of its citizens.

The subjectivity of the common law in jurisdictions of both sides of the Atlantic lends itself to further flexibility and shift in focus in the context of hostility. In the United Kingdom, the 1942 case of *Liversidge* v. *Anderson*,[17] where the House of Lords interpreted Defence Regulation 18B which allowed the Home Secretary to order a person detained if he has reasonable course to believe that such a person was of hostile origin or association. The majority decision in this case was to the effect that if the Home Secretary thinks he has good cause that was good enough. The dissenting judgment of Lord Atkin, who was of the view that judges should not be more executive minded than the executive was later upheld in the appellate stage of the Sri Lankan case *Nakkuda Ali* v. *Jayaratne*[18] where the court held that such a power, to detain persons, must be exercised on objectively reasonable grounds. In the United States, of corresponding analogy is the wartime experience where 120,000 Japanese persons were placed in detention camps during the second world war. In 1988, the United States Congress passed legislation to the effect that the prisoners had largely been detained under racial and other subjective motivation which were determinants of a weak political leadership.

The *raison de etre* of international law and the determining factor in its composition is anchored on the international political system. The domestic flavour of the *Liversidge* and *Nakkuda Ali* decisions, although admitting of the validity of internal action by a State in order to protect its internal integrity, does not lend itself to assisting the international conduct of a State, where Article 2(4) of the United Nations Charter prohibits the use of force against the territorial integrity or political independence of any State, or in any other manner inconsistent with the purposes of

[17][1942] AC 206.

[18][1951] AC 66. However, it must be noted that the Hands off the Executive approach was rekindled in the 1977 case of *Rv Secretary of State ex parte Hosenball*, a deportee case where Lord Denning said that when there was a conflict of interest between the interests of national security on the one hand and the freedom of the individual on the other, the balance between the two should be determined by the Home Secretary who is entrusted this power by Parliament. See [1977] IWLR 766 at 783.

the United Nations. Furthermore, the 1965 *Declaration of Inadmissibility of Intervention in the Domestic Affairs of States* stressed that:

> No State has the right to intervene, directly or indirectly, for any reason whatever, in the internal or external affairs of any other State. Consequently, armed intervention and all other forms of interference or attempted threats against the personality of the State or against its political, economic and cultural elements, are condemned.

This was reaffirmed in 1970 in the *Declaration on Principles in International Law*[19] where it was emphasized that not only were such manifestations abhorred by the international community, but also they were deemed to be violations of international law. Thus, there is clear direction for international lawyers to distinguish between private morality of certain authorities and public morality which is grounded upon roots of common morality and make sure that the gap between the two does not remain unbridged. In other words, governments in their conduct of foreign affairs should not be deemed as being above the law. At the same time, they should not be fettered by an unnecessarily cumbersome bureaucracy when it comes to protecting their interests and the lives of their citizens.

The enforcement of international law is strictly the purview of the national states and each state claims sovereignty to the extent that it is its own source of authority and power. In this sense, international law has no overall application on a common basis where each state can be held responsible for the adherence to a unified set of mandatory rules that can be set and enforced by one supreme legislative body. On a juridical basis however, this primitive antithesis does not leave the world totally destitute of hope. It is now very apparent that with all its inadequacies, international law is at least an entity whose presence is felt.

If one were to doubt the significance of this law, one need only imagine a world in which it were absent . . . There would be no security of nations or stability of governments; territory and air space would not be respected; vessels could navigate only at their constant peril; property – within or without any given territory – would be subject to arbitrary seizure; persons would have no protection of law or of diplomacy; agreements would not be made or observed; diplomatic relations would end; international trade would cease; international organizations and arrangements would disappear.

Logically speaking, it would therefore seem that international law establishes ethical and moral dimensions that have proven to be accepted and followed by the United Nations member States. A further indication that principles of international law have been accepted may be found in those instances in which an injured state resorts to self-help in response to branches of international law. For example, when its airspace is encroached upon, the state may request a forced landing of the offending aircraft in its territory or guide the aircraft out of its territory. The absence of organized sanctions does not create a hiatus that cannot be bridged.

The personality and character of law and the possibility of it being followed hinge firstly on the personality of the leadership from which the law emanates.

---

[19] General Assembly Resolution 2625 (XXV).

Therefore it is appropriate that a leadership of the stature of the United Nations endeavours to strengthen international law by its Resolution 44/23.

The extent to which the stature of international law has been enhanced by the United Nations is evident in yet another legal and political phenomenon – the international sanction. An essential corollary of international law and the moral and ethical fibres of the Charter of the United Nations, this measure, like the previously discussed measure of self help is now widely used by one nation against another. The reasons for application of sanctions is often the same – as retaliation against the action of one State calculated to breach provisions of the United Nations Charter. The use of self help and the imposition of sanctions are two clear and forceful instances in which both the force and the acceptability of international law are recognized by States. Although some still believe that sanctions are not effective tools of statecraft that would result in punitive action with anticipated results, there is now strong contrary opinion that the application of sanctions does effectively enhance the principles of international comity. While some view economic sanctions as mere tools that are used to achieve foreign policy goals, the fact remains that, in the ultimate analysis, a sanction effectively punishes an offending State and to that extent supplements the punitive element in international law.

The goal of a just and lasting peace can only be achieved through adherence to a set of mutually acceptable rules of conduct. If this fact was apparent after the first 25 years of the existence of the United Nations it is now even more prominently seen in the statement of former Secretary General Perez de Cuellar to the effect that after 40 years we have, for the first time in history, a virtually universal world organization. We have also for the first time in history, a world of independent sovereign States... We have achieved unprecedented economic growth and social progress... We are making collective efforts to the new generation of global problems.

International Politics

An international political system has identifiable boundaries. They pertain to geographic, cultural and environmental concerns. However, when these three parameters are coalesced, they form an anarchical society so named in contradiction to depict and represent a system which has no common power.[20] Irrespective of whether the community of States have a common power or authority, they still remain an international society with common interests and common values sharing in the working of common institutions. States recognize one another's claims to independence and sovereignty and recognize circumscriptions of one another's ability to exercise force against each other. Of course at the same time, an essential feature of international politics is the collaboration of States in sharing resources and helping each other. The prerequisites of an international society are common

[20] Bull (1977), p. 13.

values, common interests and common rules. The shift in focus from tradition international law rules to a social process transcends from a society of States to a society of humankind where sectoral boundaries would be blurred and a globalized perception of human rights and values would emerge. In such a system, individuals from all parts of the world could consider themselves an integral part of the global system, with common interests, values and rules applying to all across the globe.

Principles of international law and legal rules provide an efficient means of measuring how a certain political structure can change the identity of actors. To obtain full results, international relations and international law should be linked, particularly through scholars of both disciplines who understand the principles of each other's disciplines. Firstly, as Hurrell observes:

> it is international law that provides the essential bridge between the procedural rules of the game and the structural principles that specify how the game of power and interests is defined and how the identity of the players is established ... international law provides a framework for understanding the processes by which rules and norms are constituted and a sense of obligation engendered in the minds of policy makers.[21]

Secondly, international law and international politics occupy the same conceptual space. These two disciplines supplement each other in providing the rules and regulations for the international system, which is an intellectual construct that international lawyers, political theorists and policy makers describe as a composite whole governing the entire spectrum of international policies. It makes little sense, therefore, to study one without gaining an understanding of the other.

It can be said with some justification that international law is the thread which runs through the fabric of international politics and provides the latter with its abiding moral and ethical flavour. Without principles and practices of international law, foreign policy would be rendered destitute of its sense of cooperation and become dependent on a national self interest. As President Woodrow Wilson once claimed:

> It is a very perilous thing to determine the foreign policy of a nation in the terms of material interests ... we dare not turn from the principle that morality and not expediency is the thing that must guide us, and that we will never condone equity because it is convenient to do so.[22]

This statement, made in 1950, has great relevance today, when continued progress is being made in technological and economic development and policy decisions of States have far reaching consequences on a trans-boundary basis. Nation States are becoming more interdependent, making decisions made by a particular State in its own interest have a significant negative impact on the interests of other States. Therefore ethics in foreign policy has largely become a construct which combines cultural, psychological and ideological value structures. Within this somewhat complex web of interests, decisions have to be made, which, as recent events in history have shown, require a certain spontaneity from the international

---

[21]Hurrell (1992), p. 73.

[22]Quoted in Morganthau and Kenneth (1950), p. 24.

community. For example, when Iraq invaded Kuwait in 1990, the members of the United Nations chose economic sanctions against Iraq, claiming that war was the last resort to be embarked upon against Iraq if economic sanctions did not prove to have any effect. In hindsight, one could argue one way or another, firstly, as did the United States, that the use of force bore quick results and, on the other hand, as did many officials in Paris, Moscow, Ottawa and Washington, that the decision to wage war against Iraq was too precipitous as not enough time had been given to economic sanctions to compel Iraq to retreat from Kuwait. The precipitous but quick action taken in going to war with Iraq might be justified by some with the analogy of Britain appeasing Hitler in the 1930s without adopting a more aggressive and perhaps belligerent attitude toward German atrocities. This action, which was later labeled as folly by most political scientists, was applauded and endorsed at that time in the British Parliament.

For years the construct of international politics and foreign relations has been that we live in a global village. The ultimate reality of this concept was seen on 11 September 2001 when the world intimately shared the tragic disasters of the United States. The United States, which had policed the world for 56 years after World War II, lost its innocence and its isolation, suddenly turning into a different kind of world power to whose aid other nations rushed. Global cooperation reached its zenith, with NATO invoking Article 5 of its founding treaty and declaring that terrorist activity against one NATO member country constituted an attack on all NATO member States, which would collectively respond as if they had been attacked themselves. The President of Russia phoned the President of the United States and pledged Russia's support. A week after the attacks, on 19 September 2001, the Organization of American States invoked their mutual defense pact – the Rio Treaty. The coming together of the entire world (except Iraq) in support of the United States signaled the dawn of a new cause and a new sense of challenge. New partnerships between governments were forged toward attaining a new global coalition against terrorism.

States come together as one in times of crisis and remain with each other at other times mainly because international relations have been a rule based system. Therefore, deviating from established rules of international law would act to the detriment of international unity and global cooperation among States. The first step toward ensuring a cohesive international consensus system is to continue building a rule based, cooperative democratic global system which does not succumb to individual acts of terror or conflict. Such a global system would not only ensure peace among nations but would also confront injustice, poverty, disease and environmental hazards.

The world must heed the call of some States which are now appealing to upkeep the principle of responsibility to protect which is enshrined in the United Nations Charter as the basic mandate of the United Nations. This mandate is primarily of the Security Council which was designed to meet threats and dangers posed by cross-border aggressions. States must not be allowed to indulge in preventive action in a unilateral manner, merely because there is a perceived security risk or threat to their peaceful existence.

Restoration of the dignity of the United Nations Security Council should go hand in hand with increased collaboration and partnership between States, through the United Nations, to create institutions that could address burning issues relating to internationally displaced persons, refugees and poverty. Global disarmament is another problem to be tackled. The proliferation of arms around the world has spurned threats not only to international harmony but also to domestic peace within nations. A more democratic system than that which exists in the United Nations system at present could help in developing principles of international intervention when States concerned or affected are not prepared or willing to take measures in protecting their own people. The United Nations should be geared more toward safeguarding people's interests through an international judicial system that would supplement the International Court of Justice which hears disputes between States and gives advisory opinions on issues regarding States. If the shifting focus of international law and politics would finally beam on the human being rather than the collective interests of the States, and human rights are ensured through various existing United Nations institutions and new ones which collect, disseminate and use information on abuse of human rights with a view to correcting them, such a shift of focus would indeed be a good thing.

The above recommendations, when applied to air and space law, have special meaning, as the issues addressed in this book will reflect. More emphasis is placed on the one hand on the interests of States whilst on the other, the rights of the individual are not neglected. These rights have to be viewed in the context of a changing world, where security and expediency of travel have to be considered as being symbiotic.

Air Space and Outer Space

The issue of air space and outer space is looming over the aerospace community, particularly with the prospect of space travel on a commercial basis being already a reality. At the time of writing, the aerospace community was considering such issues as sub-orbital flights and space tourism, both of which could further blur the boundaries between air space and outer space, while raising issues of topical interest. So far, there has not been a universally accepted definition distinguishing air space and outer space. Some years ago, when the legalities of an aerospace plane, which is a hypersonic single stage to orbit reusable vehicle that horizontally takes off and lands on a conventional runway were considered, it was thought that the transit through near space which is involved is incidental to the main transit which takes place within the airspace. Generally, the aerospace plane, which will be constructed with the use of aeronautical and space technologies and would be capable, and, indeed, required to fly both in airspace and outer space, would bring to bear the need to consider the applicability of and appropriateness of laws relating to the space plane's activities. It will be subject to the sovereignty of the State whose airspace it is in. This is an incontrovertible fact which need not be stated since any object within the airspace of a territorial State would indeed be subject to that State's sovereignty.

At an open forum discussion held on 6 April 1989 during the Annual Meeting of the American Society of International law in Chicago the legal and policy issues of the aerospace plane were brought to focus, and on the subject of delimitation of applicable laws, a view was expressed by academics of McGill University's Institute of Air and Space Law that the activities of the aerospace plane as a space traversing device used for point to point earth transportation could be governed by air law. It was the view of these academics that should the principle of air law apply, it was necessary for bilateral agreements to be signed between States prior to an international flight by the aerospace plane. It is logical to assume that, speaking from a purely commercial point of view, it follows that the aerospace plane which transports persons and goods between States should be considered an aircraft, even if it traverses through space at a given time.

Recently, the official launch of space tourism, where paying customers travelled beyond Earth's atmosphere, gave rise to an entirely different dimension, where the different issue of sub-orbital flights has emerged as requiring some consideration, particularly on the question as to whether such flights travel to outer space or whether they are deemed to be considered as not leaving the Earth's atmosphere. Unlike the aerospace plane which would leave the territory of one State as an aircraft, enter outer space and travel in outer space until it descends to a destination State, sub-orbital flights would not usually travel between two States but would ascend to an altitude sufficient for the persons on board to view the Earth as a whole globe, a phenomenon not available to aircraft passengers. The vehicle would descent to the State from which it took off. This activity is called "sub-orbital flying" and is gaining increasing popularity in the realm of space tourism. One of the issues that sub orbital flights raise is whether, at the height the flights are conducted, the vehicle is deemed to be in air space or outer space. Therefore, sub orbital flights inevitably call for a determination as to what might be air space, as against outer space.

The Permanent Court of International Justice, when requested for a definition of "air space" in the 1933 *Eastern Greenland's Case*,[23] was of the view that the natural meaning of the term was its geographical meaning. The most fundamental assumption that one could reach from this conclusion is that air space is essentially geophysical, meaning that it is space where air is found. Simplistically put, "air space" has been considered as going upwards into space from the territorial boundaries of a State and downwards to the centre of the Earth, in the shape of an inverted cone. This theory, advanced mathematically, in terms of space where air is found, would encompass the atmosphere, which has is layered into components starting from the troposphere (from sea level to about 10 kilometres); the stratosphere (from about 10 to 40 kilometres up); the ionosphere (from about 40 to 375 kilometres); and the exosphere (from 375 to 20,000 kilometres). Based on this methodology, a

[23]PCIJ Series A/B, No. 53, at pp. 53ff.

sub-orbital flight, which goes up to about 62.5 miles (100 kilometres) above the landmass of the Earth, would hover somewhere in the lower level of the ionosphere, prompting the conclusion that it is a space flight traversing outer space, while others would maintain that the vehicle does not leave the Earth's atmosphere and therefore is airborne.

The United Nations Committee on the Peaceful Uses of Outer Space (UNCOPUOS), which is the UN forum where technical and legal aspects of space activities with global impact are considered, has discussed the issue of the definition and delimitation of outer space from 1962 and no definite conclusion has been reached so far in this regard. In this connection, it is of interest to note that the Legal Subcommittee of UNCOPUOS, through its Working Group on Matters Relating to the Definition and Delimitation of Outer Space, has been considering possible legal issues with regard to aerospace objects. A questionnaire thereon has been circulated to all U.N. Member States. A compilation of the replies received so far and an analytical summary of such replies, as well as a historical summary on the consideration of the question on the definition and delimitation of outer space, may be found on the OOSA website.[24]

As debated for decades in the framework of UNCOPUOS, it may be questioned whether the vertical limit of airspace would be critical to determine the scope of applicability of air law as opposed to international space law conventions (spatialist approach), or whether the type of activities at issue would determine which law should apply (functionalist approach) to sub orbital flights. The latter school of thought submits that flights which would be passing merely in transit through (sub) orbital space in the course of an earth-to-earth transportation would be in air space and therefore remain subject to principles of air law.

A sub-orbital flight is a flight up to a very high altitude which does not involve sending the vehicle into orbit. 'suborbital trajectory', which a sub orbital flight would follow, is defined in the legislation of the United States as "The intentional flight path of a launch vehicle, re-entry vehicle, or any portion thereof, whose vacuum instantaneous impact point does not leave the surface of the Earth."

In 2004, SpaceShipOne was the first private vehicle to complete two sub-orbital flights within two weeks carrying weight equivalent to three human adults up to about 62.5 miles (100 km) to win the Ansari X Prize. It was carried during one hour by an aeroplane up to nearly 50 000 feet (9.5 miles) from where it was released into a glide and then propelled vertically for 80 seconds by a rocket motor to an altitude of more than 62 miles at apogee, reaching a speed over Mach 3. Then falling back to return to earth, it re-entered the atmosphere and glided during 15 to 20 minutes before landing back on the runway of departure.

SpaceShipOne, strictly speaking, does not operate as an aeroplane or even as an aircraft during the ballistic portion of the flight while it is not supported by the reactions of the air, even though some degree of aerodynamic control exists throughout the trajectory from launch altitude until the craft enters the upper

[24] www.oosa.unvienna.org/index.html.

reaches of the atmosphere where the air density is no longer sufficient for aerodynamic flight. After apogee, during re-entry into the atmosphere the vehicle transitions to unpowered aerodynamic (gliding) flight for the return to earth. Consequently, depending upon some design and operational aspects, it could be considered operating as an aircraft in flight during this latter portion of the journey.

Therefore, such vehicles could fulfil the principal elements in the definition of aircraft and be used as such during a portion of their flights, but they offer some characteristics of a rocket as well. It is likely that other vehicles engaged in the future in such sub-orbital flights would similarly be of an hybrid nature, taking into account that developments to come may lead to a range of designs, some of which could be more clearly classified as aircraft. Should sub-orbital vehicles be considered (primarily) as aircraft, when engaged in international air navigation, consequences would follow under the Chicago Convention, mainly in terms of registration, airworthiness certification, pilot licensing and operational requirements (unless they are otherwise classified as State aircraft under Article 3 of the Convention).

Plans have been announced by Virgin Galactic for the development of a fleet of five sub-orbital vehicles to carry paying passengers, six per vehicle; it plans that the first of these will be ready for commercial operations in 2008 at the earliest. There are indications that at least one other company is planning to offer rival sub-orbital flights.

Manned and unmanned sub-orbital flights have been undertaken to test spacecraft and launch vehicles intended for later orbital flight, but some vehicles have been designed exclusively to reach space sub-orbitally: manned vehicles such as the X-15 and SpaceShipOne, and unmanned ones such as ICBMs and sounding rockets. Sub-orbital tourist flights will initially focus on attaining the altitude required to qualify as reaching space. The flight path will probably be either vertical or very steep, with the spacecraft landing back at its take-off site.

The spacecraft will probably shut off its engines well before reaching maximum altitude, and then coast up to its highest point. During a few minutes, from the point when the engines are shut off to the point where the craft begins to slow its descent for landing, the passengers will experience.

A suborbital flight is known to be the next generation of commercial passenger travel. At the present time flight testing of commercial reusable launch vehicles (RLVs) is underway, making the availability of frequent suborbital flight closer than ever. As earlier mentioned sub orbital flights fly out of the atmosphere but do not reach speeds needed to sustain continuous orbiting of the earth. They allow passengers to look down at the brilliant curvature of the earth as they would from orbit.

One must not confuse a sub orbital flight with a space flight which is a flight *into* or *through* space. The craft which undertakes a spaceflight is called a spacecraft. It is often thought that orbital spaceflights are spaceflights and sub-orbital spaceflights are less than actual spaceflights. This is not entirely accurate as both orbital and sub-orbital spaceflights are true spaceflights.

The term *orbit* can be used in two ways: it can mean a trajectory in general, or it can mean a closed trajectory. The terms *sub-orbital* and *orbital spaceflights* refer to the latter: an orbital spaceflight is one which completes an orbit fully around the central body.

From the above discussion the conclusions that could be drawn are that for a flight from Earth to be a spaceflight, the spacecraft has to ascend from Earth and at the very least go past the edge of space. The edge of space is, for the purpose of space flight, often accepted to lie at a height of 100 km (62 miles) above mean sea level. Any flight that goes higher than that is by definition a spaceflight. Although space begins where the Earth's atmosphere ends, the atmosphere fades out gradually so the precise boundary is difficult to ascertain. Therefore one could argue that there is a need to accept the fact that vehicles which would effect earth-to-earth connections through sub-orbital space could incorporate the constitutive elements of aircraft and fly as such at least during descending phase while gliding. However, rocket-propelled vehicles could be considered as not falling under the classification of aircraft.

From a spatialist viewpoint, there is no clear indication in international law on the delimitation between airspace and outer space which would permit to conclude on the applicability of either air law or space law to sub-orbital flights. On the other hand, it might be argued from a functionalist viewpoint that air law would prevail since airspace would be the main centre of activities of sub-orbital vehicles in the course of an earth-to-earth transportation, any crossing of outer space being brief and only incidental to the flight. The United Nations Committee on the Peaceful Uses of Outer Space (UNCOPUOS), and more particularly its Legal Subcommittee, is considering the question of possible legal issues with regard to aerospace objects but no final conclusion has been reached yet.

References

Arend AC (1999) Legal rules and international society. Oxford University Press, Oxford, 26
Bull H (1977) The Anarchical Society. Yale University Press, New Haven, CT, 128
Freidman W (1964) The changing structure of International Law. Butterworths, London, 213
Higgins R (1999) Problems and process: international law and how we use it. Oxford University Press, Oxford, 1
Hurrell A (1992) International society and the study of regimes: a reflective approach. In: Rittberger V (ed) Regime theory and international relations, vol 49. Butterworths, Sydney, p 73
Jennings RY (1990) An international lawyer takes stock. Int Comp Law Q 39(3):513
Johnson D (1955–1956) The effect of the resolutions of the General Assembly of the United Nations. Br Yearb Int Law 32:97
Kennedy D (1988) A new stream of international law scholarship. Wis Int Law J 7(1):8
Lowe V (2003) The Iraq crisis, what now? Int Comp Law Q 52(4):859–871
Morganthau H, Kenneth W (1950) Thompson, principles and problems of international politics. Knopf, New York, 24
Purvis N (1991) Critical legal studies in International Law. Harv Int Law J 32(81):115
Shaw MN (2003) International law, 5th edn. Cambridge University Press, Cambridge, 1

Chapter 2
Space Security

Space security has been defined as *the secure and sustainable access to, and use of, space and freedom from space-based threats.*[1] The view has been expressed that:

> This working definition provides a vision of space security that has proven to be useful in guiding our collective consideration of space security relevant issues. Space actors have thus far been able to access and use space for a wide variety of civil and military applications without having significant negative impacts on the access to and use of space by others, or without exporting terrestrial conflicts into space. This definition also served as an important point of reference in order to seriously address emerging concerns regarding space security, including those related to a growing dependency upon space assets for critical national security functions combined with a growing sense of the vulnerability of these same assets.[2]

There are nine indicators of space security: the space environment; space situational awareness; space security laws, policies and doctrines; civil space programs and global utilities; commercial space; space support for terrestrial military operations; space systems protection; space systems negation; and space-based strike weapons. The *Space Security Index* of 2010 distinguishes between militarization and weaponization of space:

> An important distinction must be made between militarization and weaponization of space: while the former is a reality, thus far there is no documented evidence of the latter. Although the use of space assets for military applications such as reconnaissance, intelligence, and troop support has been ubiquitous for several years, space apparently has remained weapons-free. To maintain this state, the prevention of an arms race in outer space remains a priority for policymakers at various international forums, since it is assumed that once a state places weapons in space, others will follow suit.[3]

The United States and the Union of Soviet Socialist Republics (USSR) developed a wide range of extensive military space systems during the Cold War. In this context they used satellites which offered an ideal vantage point from which to monitor the Earth. They were also used as harbingers of danger from outer space

---

[1] Lawson (2004), p. 177.

[2] Lawson (2004), p. 177.

[3] http://www.spacesecurity.org/publications.htm.

R. Abeyratne, *Space Security Law*,
DOI 10.1007/978-3-642-16702-7_2, © Springer-Verlag Berlin Heidelberg 2011

that could provide strategic warning of signs of nuclear attack, such as the launch plume of a ballistic missile or the light signature of a nuclear detonation. Arms control verification was another reliable medium offered by satellites, which allayed the fears of the United States in particular, that the USSR could have a strategic advantage over the United States in missile weaponry. The space age, which commenced in the 1950s, opened new vistas in the development of reconnaissance, surveillance, and intelligence collection capabilities through the use of satellite imagery and space-based electronic intelligence collection. One of the most useful uses of satellite imagery was the capabilities offered in the command and In addition, satellite communications provided extraordinary new control of military forces deployed throughout the world. It is worthy of note that over fifty percent of all space communications systems have been developed to support military activity on Earth and this makes the military aspect of space exploration and technology an important consideration in national security, politics and international relations. The proliferation of activity, leading to an increased use of outer space has started a race for orbital slots and allocations for the use of the radio spectrum.

Satellite imagery, and in particular remote sensing, which will be addressed later in this book, introduces a dimension of caution in the area of national security and privacy of States. The Report of the 72nd Conference of the International Law Association, held in Toronto in June 2006, makes the pertinent observation that the stark dichotomy between sensing and sensed States is being blurred bringing to bear new opportunities in the further development of remote sensing law. Sensing States are usually the developed States and sensed States are usually developing States. The new development is because some developing countries are now becoming sensing countries which brings out the need to develop State practices with regard to remote sensing. At space law, the role between sensing and sensed States is linked by equity, which, usually is a bridge between man and law but is extended to outer space, as originally recognized by the space law pioneer Justice Manfred Lachs.

Remote sensing is a "catch all" expression which describes a set of technologies designed to collect information relating to the physical properties of the Earth. The legal principles of remote sensing are enshrined in United Nations General Assembly Resolution 41/65, adopted in December 1986 at the Assembly's 41st Session. The term "remote sensing," according to the Resolution, means the sensing of the Earth's surface from space by making use of the properties of electromagnetic waves emitted, reflected or diffracted by the sensed objects, for the purpose of improving natural resources management, land use and the protection of the environment; Popularly called "eye in the sky," remote sensing assists diverse fields of human interaction, from meteorological services in forecasting weather to criminologists' work in re-creating scenes of accidents and crimes. UN Resolution 41/65 also provides that remote sensing activities shall be carried out for the benefit and in the interests of all countries, irrespective of their degree of economic, social or scientific and technological development, and taking into particular consideration the needs of the developing countries. The dynamic applications of a spatial satellite imaging system will be invaluable to attorneys, insurers and risk managers in acquiring data and information on natural and man-made disasters. The Chernobyl Nuclear Plant disaster and the Exxon Valdez oil spill in Alaska are two such incidents which were tracked and recorded for use by television and major newspaper media. The most significant area of contribution is in prediction or preparation where, a satellite just overhead of a disaster area can identify through its imagery an impending disaster or occurrence. Currently remote sensing by satellite is being carried out by several countries around the world, such as USA, Russian Federation, France, India, China, Israel, Japan, Brazil and Argentina.

Remote sensing is accomplished through satellite imagery, which involves a process which uses cameras or sensor systems usually mounted on an orbiting satellite that capture light reflected from the earth's surface. The fundamental principle of this imagery process is that natural and man-made materials absorb and reflect varying quantities of light in different wavelengths and, through this absorption and reflection process, energy (light) enables satellite sensors to collect data by interacting with objects on the surface of the earth such as plants, soils and buildings in a way that makes possible the extraction of such data and information. Such information is absorbed by the camera/sensor system on the orbiting satellite and transmitted back to earth in a digital format and transformed or converted to images that are capable of being interpreted by sophisticated image processing software. This process makes satellite imagery ideally applicable to the requirement for detail pertaining to large area coverage such as regional cartography, environmental assessment, infrastructure planning and agricultural monitoring.

One of the seminal purposes of remote sensing is to promote the protection of the Earth's natural environment. To this end, States participating in remote sensing activities that have identified information in their possession that can be used to avert any phenomenon harmful to the Earth's natural environment are required to disclose such information to States concerned.

Another purpose of remote sensing is to promote the protection of mankind from natural disasters. To this end, States participating in remote sensing activities that have identified processed data and analysed information in their possession that may be useful to States affected by natural disasters, or likely to be affected by impending natural disasters, are required to transmit such data and information to States concerned as promptly as possible.

It is also required at international law that, as soon as the primary data and the processed data concerning the territory under its jurisdiction are produced, the sensed State will have access to them on a non-discriminatory basis and on reasonable cost terms. The sensed State shall also have access to the available analysed information concerning the territory under its jurisdiction in the possession of any State participating in remote sensing activities on the same basis and terms, particular regard being given to the needs and interests of the developing countries. Furthermore, in order to promote and intensify international cooperation, especially with regard to the needs of developing countries, a State carrying out remote sensing of the Earth from space is required, upon request, to enter into consultations with a State whose territory is sensed in order to make available opportunities for participation and enhance the mutual benefits to be derived therefrom.

Finally, it is recognized that States operating remote sensing satellites will bear international responsibility for their activities and assure that such activities are conducted in accordance with the provisions of the Treaty and the norms of international law, irrespective of whether such activities are carried out by governmental or non-governmental entities or through international organizations to which such States are parties. This principle is without prejudice to the applicability of the norms of international law on State responsibility for remote sensing activities. The first area of State liability that comes to mind involves the use of defective outer space equipment that would give incorrect information to air navigation services. Liability in this instance could ensue if damage to persons or property is caused by the use of such information.

The refusal of a State which holds data obtained from remote sensing of the Earth's surface to provide such information to another State may result in a claim by the latter State that the withholding of such data is an abuse of its rights and that such refusal is calculated to result in such information being used against the requesting State. The aggrieved State may even claim that the State which holds the data would use the information to its advantage, resulting in the unjust enrichment of the State which holds the technology and information at the expense of the other.

Although it is an entrenched principle at space law that sensed States have a right of access to both the raw and processed data concerning their territories, which is an unqualified right of access, there is an increasing tendency among sensing States to deny that right of access to a State whose territory is sensed by them.

Another factor that is noteworthy is that remote sensing of the Earth's surface is an issue that opens itself to State liability issues. The notion of using devices such as satellites to survey the globe, although incontrovertibly beneficial to the development of cartography, meteorology, geography and oceanography – all of which affect air navigation – would inevitably raise questions of State privacy. Some may even argue that the sharing of satellite resource data may derogate Article 2(7) of the United Nations Charter which prohibits State intervention in matters which are essentially within the domestic jurisdiction of any State. The unauthorized space observation of a territory of a State by another has been identified as a deprivation of a right of the observed State, even if no more information is obtained than that which is already available to that State, resulting in the award of pecuniary compensation.

A significant development in recent satellite imagery through spatial information is the launching of the highest resolution commercial remote sensing satellite with an on-board digital camera having the capability of producing one-metre or better resolution at nadir to focus on the lowest point in the panchromatic, full spectrum mode and four-metre resolution imagery in the red, green, blue and near infra-red (NIR) bands. Such a one-metre resolution will enable the user to distinguish between objects which are one meter in size on the ground if they have distinguishable physical and visual characteristics. Easily detectable under this spatial recognition system are stripes in parking lots, swimming pools, cars, trucks, boats, tennis courts, landscape features, all within their surroundings and environs.

Satellite imagery is not restricted to photography. New sophisticated electronic sensors enable even more powerful imagery that mere photos – digital data that can be analysed, processed and interpreted on computers – computer analysis in digital format admits of possibilities for data to be used in multifarious ways using differences in spectral responses from ground features. The usefulness of this process is emphasized and highlighted when satellite imagery is used in conjunction with the geographic information system (GIS) and the global positioning system (GPS), which together can construct a complete model of an area.

United Nations General Assembly Resolution 41/65 provides that Remote sensing activities shall be conducted in accordance with international law, including the Charter of the United Nations, the Treaty on Principles Governing the Activities of States in the Exploration and Use of Outer Space, including the Moon and Other Celestial Bodies, and the relevant instruments of the International Telecommunication Union. The Resolution, in Principle 5 requires that remote sensing activities shall be conducted in accordance with the principles contained in article I of the Outer Space Treaty, including the Moon and Other Celestial Bodies, which, in particular provides that the exploration and use of outer space shall be carried out for the benefit and in the interests of all countries, irrespective of their degree of economic or scientific development, and stipulates the principle of freedom of exploration and use of outer space on the basis of equality. These activities shall be conducted on the basis of respect for the principle of full and permanent sovereignty of all States and peoples over their own wealth and natural resources, with due regard to the rights and interests, in accordance with international law, of other States and entities under their jurisdiction. Such activities shall not be conducted in a manner detrimental to the legitimate rights and interests of the sensed State.

States carrying out remote sensing activities are required to promote international cooperation in these activities. To this end, they are required to make available to other States opportunities for participation therein. Such participation shall be based in each case on equitable and mutually acceptable terms. In order to maximize the availability of benefits from remote sensing activities, States are encouraged, through agreements or other arrangements, to provide for the establishment and operation of data collecting and storage stations and processing and interpretation facilities, in particular within the framework of regional agreements or arrangements wherever feasible.

States participating in remote sensing activities are legally bound to make available technical assistance to other interested States on mutually agreed terms. From a global perspective, the United Nations and the relevant agencies within the United Nations system are required to promote international cooperation, including technical assistance and coordination in the area of remote sensing. A State carrying out a program of remote sensing has to inform the Secretary General of the United Nations. It shall, moreover, make available any other relevant information to the greatest extent feasible and practicable to any other State, particularly any developing country that is affected by the program, at its request.

The protection of knowledge and communication is another area which requires careful consideration when legal issues relating to the application of space technology to air navigation is considered. These aspects of space technology would be protected under patent laws. While ground based space technology and communications arising therefrom would be protected by patent laws applicable normally both nationally and internationally, it is arguable that new issues in the realm of patent laws relating to outer space technology may have to be addressed. A critical factor for consideration would be the specific instances where a State would withhold information from air navigational facilities where such information has been obtained through the use of space technology and could affect the security of the State which holds such information.

The sharing of knowledge through the use of space technology is incontrovertibly linked to principles of space law that would admit of such use. The basic principle of space law that addresses this issue is the "common interest" principle which emerged as a result of the first specific Resolution on space law of the United Nations General Assembly in 1958. The "common interest" principle has since been incorporated in subsequent multilateral treaties, particularly the *Outer Space Treaty* of 1967, Article 1(1) of which provides that :the exploration and use of outer space, including the moon and other celestial bodies, shall be carried out for the benefit and in the interest of all countries, irrespective of their degree of economic or scientific development, and shall be the province of all mankind.

This provision, which binds signatory States is seemingly a departure from the traditional "national interest" approach of international law and has represented a moral obligation to some, while to others the provision has represented a mandatory legal principle. Whatever be the approach taken by some States in refusing to share information and findings with other States, particularly those States that are sensed who request the sharing of such information. Since much of space exploration is conducted for the benefit of humanity, primacy must be accorded to those principles of law that entrench legal rules pertaining to human conduct and social benefit. In this perspective, principles of space law, as contained in the space law treaties discussed above support the basic principles of human welfare applicable in the field of aviation. Therefore, rules of conduct for States and other actors should be considered as being integral to the obligations of States whether or not they are enforceable at law.

The Global Navigation Satellite System

One of the landmarks in satellite technology was the United States Global Navigation Satellite System (GNSS) called the Global Positioning system (GPS) which emerged after the end of the Cold War. The Russians had also begun to develop satellite navigation systems that provided increasingly accurate geographical positioning information. The Russian GLONASS is a GNSS in the process of being restored to full operation (21 of 24 satellites are operational). The Galileo positioning system of the European Union is a GNSS in initial deployment phase, scheduled to be operational in 2014. The Peoples Republic of China has indicated it will expand its regional Beidou navigational system into the global Compass navigation system by 2020.

The United States extended the capability and application of GPS to further develop the role of military space systems, by integrating them into virtually all aspects of military operations which ranged from the provision of providing indirect strategic support to military forces to enabling the application of military force in near-real-time tactical operations through precision weapons guidance. A notable development in this regard was the introduction of radar satellites offered the potential to detect opposition forces on the ground in all weather at all times.

GNSS has been hailed as the greatest scientific revolution of the twenty first century.[4] The *au courant* technology is subset of the various applications of space technology via satellites such as telecommunications, broadcasting, remote sensing etc. The GNSS systems are a constellation of orbiting satellites that work in tandem with a well developed network of ground stations to detect and deliver high precision data regarding three dimensional position and time of the aircraft. The global aviation community at present has already savored and is astounded by the useful technology provided by GPS[5] (Global Positioning systems) and GLONASS[6] (Global Orbital Navigation Satellite system constellations that belong to the

[4]I am grateful to my colleague Usha Balasubramaniam, LLM for providing some material on GNSS and the aviation analogy that follows.

[5]GPS was originally developed by the Department of Defence, Government of United States. Since 1994 GPS employed a constellation of 24 satellites to provide positional and timing information to civilian and military users of an accuracy of less than 15 meters anywhere on earth. The Civilian uses/applications of this technology are very wide and has the propensity to shock the inventors of this novel technology. The civilian utilities involve computer guided automobile tracking systems, emergency/location based services, digital cartography, land surveying, navigation on land, sea and skies, high precision self calibrated smart watches, mobile communications, etc. and the basic of the military applications involves high precision guided missiles/munitions, atomic clocks, etc. It can be noted here that the civilian applications of GPS technology far precedes its defense related applications. In the year 2005 new generation of GPS satellites have been added to the constellation to increase its coverage and military capabilities. The civilian users use the intentionally downgraded Standard Positioning service (SPS) that has been made available free of cost and without restrictions. The Precise Positioning Service (PSP) is meant for the Department of Defense applications. The FAA website highlights the pivotal role played by GPS signals in the Air Transport industry and expresses heightened concerns about its vulnerability to intentional or unintentional jamming and Radio Frequency Interference. Other sources of concern for FAA are issues pertaining to reliability, accuracy and integrity of GPS signals bound for air navigation purposes. On this front FAA developed WAAS (Wide Area Argumentation System) and LAAS (Local Area Argumentation System) for alerting and warning GPS users about nonreliance of GPS signals in times of trouble. FAA also seeks to enhance the accuracy of GPS signals critical to air navigation through the DGPS initiative by minimizing position errors within the range of 10 meters.

[6]GLONASS the Russian variant of GPS provides equally good, reliable time and position related signals and is owned and operated by the Russian government. GLONASS is set on a constellation of 24 satellites which became fully operational in 1995. In terms of the precision range GLONASS promises the civilian user of precision in time and position related information within the range of 100 meters and military users of precision of less than 10-20 meters. The GLONASS signals were used extensively during the Chechnya operations.

governments of United States and Russian Federation respectively.[7] The European counterpart "GALILEO"[8] is soon to join the league and "is slated to reach full operational capacity in 2008."[9]

The market for civilian uses of GPS was to grow at the rate of "22 billion in 2008, according to ABI, a New York-based technology market research firm."[10] "The US GPS Industry Council has quoted figures for worldwide GPS equipment sales to around $6bn in 1999, rising to $16.5bn for 2003."[11] "By the year 2000 civil users outnumber military users by 100 to 1 and the ratio was increasing. The Compound Annual Growth Rate of the GPS market was growing by approximately 22%."[12]

> Due to the all-embracing nature of GPS, the relatively new nature of the market, its evolving technology and the creation of augmentation services, analysts' views as to the development of the market differ considerably. There is no consistent view as to the rate of development and the degree of penetration of GPS into the consumer sector.[13]

"Europe is said to gain 62 billion Euros in terms of economics benefits and 12 billion dollars in terms of social benefits by setting up of the Galileo project."[14]

The aviation sector is one of the biggest users of GPS/GLONASS technology. Though the service per se is being made available free of direct user charges, heavy investment is expected on lines of ground and aircraft based receivers and related support infrastructure to make use of this technology for navigation. Further, it should be remembered that if United States and Russian Federation discontinue

---

[7]GALILEO: The European counterpart is all set to start its operations from the year 2008. It promises the civilian users accurate, reliable data on par with the military users. The system was developed keeping in view the disadvantageous position of Europe in the GNSS systems market, reduced accuracy of GPS data and doubtful reliance on GPS/GLONASS operated by defense regimes of US and Russian Federation. The selective availability (A program disbanded by US Government since 2000) of GPS during Bosnian crisis immediately triggered a need for an independent, self-reliant GNSS technology in Europe. Galileo promises precision, accurate data with one meter range.

[8]Exchange of letters between ICAO and US and Russian Government in October 1994 and July 1996 for making GPS and GLONASS available to the global aviation community for a period of at least 10 years and 15 years respectively with no direct user charges.

[9]What is GNSS? available at http://searchnetworking.techtarget.com/sDefinition/0,,sid7_gci827051,00.html, accessed on 12/08/2006.

[10]Leonard David, Satellite Navigation: GPS grows up, Market booms', online information available at http://www.space.com/businesstechnology/technology/satcom_gps_overview_031105.html, accessed on 12/08/2006.

[11]Global Positioning System Report published by Ireland based Research and Markets, Online information available at http://www.researchandmarkets.com/reports/3687/3687.htm, accessed on 12/08/2006.

[12]NAVSTAR, available at http://www.astronautix.com/project/navstar.htm, accessed on 13/08/2006.

[13]NAVSTAR, available at http://www.astronautix.com/project/navstar.htm, accessed on 13/08/2006.

[14]Lammerstma (2005), p. 14.

making GNSS technology available free of cost on the termination of the minimum mandatory period as per the July, 1994 and September, 1996 letters of exchange with ICAO and make it available on cost recovery basis[15] some additional billions would roll into billion dollar GPS/GLONASS/GALILEO markets.

The statistics above depict the commercial scope in terms of market potential of "Aviation GNSS projects." Looking at GNSS from cost point of view will throw light on how expensive establishment and maintenance of GNSS systems could be? The statistics which follow are a grim reminder of the investment realities surrounding the GNSS drive.

- "It takes US Government approximately 400 US dollars every year for full operability and maintenance of GPS."[16]
- "The Russian Government experienced budgetary constraints and has join hands with Indian Government to redeploy 18 of the satellites again."[17]
- "The cost of setting up Galileo would be 3.5 billion Euros and its operation costs is said to be 2.5 billion Euros."[18]

The substantial investment of the nature mentioned above is a major entry barrier and main reason for restricted entry into the GNSS market making it another ripe venue of confrontation between the GNSS capability haves and have nots. Many nations of the world might not have the means to set up and maintain such independent facilities though its in their interest to do so. This technology is in the hands of a few and the rest of world has to depend on these countries. This is the major Achilles heels of the CNS/ATM project. The commercial interests in making available these services would far exceed other considerations other than war and security emergency ensuring it's accurate and continuous availability. With the development of new navigation satellite systems by other countries such as the European Union's Galileo, Japan's Quasi-Zenith Satellite System, and China's Twinstar[19] and consequent reduction of increased reliance on the traditional core satellite systems of GPS and GLONASS guarantees a bright future for ICAO's CNS/ATM by increasing the prospect of more states willingly implementing this project.

The legal viability of the CNS/ATM system has continuously piqued the international and national policy makers from the time of conception of this novel project. At all stages of development and implementation of the project apprehensions on the legal front were repetitively expressed on lack of a long-term liability

---

[15]Article 15 of the Chicago Convention is applicable to the charges pertaining to GNSS services, Article 5 of Draft Contractual Framework relating to the Provision of GNSS Services (SSG-CNS/8-Report) and Article 5 of A32-19 'Charter on Rights and Obligations of States relating to GNSS services'.

[16]Information on GPS available at http://en.wikipedia.org/wiki/Global_Positioning_System, accessed on 14/08/2006.

[17]GLONASS, available at http://en.wikipedia.org/wiki/GLONASS, accessed on 13/08/2006.

[18]Lammerstma (2005), p. 14.

[19]Kolodziej (2006).

regime, which were discussed elaborately on the following occasions, which are as follows[20]:

- Report of the First, Second and Third Meetings of the Panel of Legal and Technical Experts on the Establishment of Legal Framework with regard to GNSS held in the years 1996, 1997 and 1998 respectively.
- Resolution A 32-19 on Charter on the Rights and Obligations of states Relating to GNSS Services.
- Resolution A 32-20 on Development and elaboration of an appropriate long term legal framework to govern the implementation of GNSS.
- CNS/ATM Implementation Conference in Rio de Janeiro in May, 1998.[21]

"A consensus has been reached that there are no legal obstacles to implementation of CNS/ATM projects and that it is completely in sync with Chicago Convention and its annexes."[22] "Two General Assemblies have promulgated resolutions declaring the organization's commitment to GNSS and that there are no legal obstacles to its implementation and that GNSS project is completely compatible to Chicago Convention and international law"[23]and further an international framework was established for Rights and Obligations of states relating to GNSS services. But lack of a multilateral framework establishing liability principles in case accidents result from errors emanating from the complex technical GNSS arrangement has been a main cause for resentment amongst the nations to adopt the GNSS. Chicago Convention is not a liability based regime, thereby it cannot come to the rescue in the issue above. "The present liability systems have been criticized for being inconsistent and inefficient, but those perceived weaknesses substantially predate GNSS."[24]

Resolution A 32-10 on "Development and Elaboration of an appropriate long-term legal framework to govern the implementation of GNSS" instructed the ICAO Council and the Secretary General within their respective competencies, beginning with a Secretariat Study Group to:

a) Ensure the expeditious follow-up of the recommendations of the worldwide CNS/ATM systems Implementation Conference, as well as those formulated by LTEP, especially those concerning institutional issues and questions of liability and
b) Consider elaboration of appropriate long-term legal framework to govern the operation of GNSS systems, including consideration of an International

[20]Legal Issues regarding Global Navigation Plan for CNS/ATM Systems, Chapter 11, Document 9750 AN/963, Second Edition of 2002, at p. 1-11-1.

[21]Legal Matters, Part V, Assembly Resolutions in Force (as of October 2004), Document 9848 at v-4.

[22]Legal Issues regarding Global Navigation Plan for CNS/ATM Systems, Chapter 11, Document 9750 AN/963, Second Edition of 2002, at p. 1-11-1.

[23]Micheal B. Jennison, A Legal Framework for CNS/ATM, in his speech on 14th November 1998.

[24]Micheal B. Jennison, A Legal Framework for CNS/ATM, in his speech on 14th November 1998.

convention for this purpose, and to present proposals for such a framework in time for their consideration by the next ordinary Sessions of the Assembly

Formulation of a liability regime for CNS/ATM project seemed to be an arduous task because this technology infrastructure involves ground as well as space segments. Some experts had pointed out during the Second LTEP meeting that "the issue of liability here was a complex one, because it involved both air law and space law and the applicability of space law treaties were emphasized, particularly in case of failure of the signal in space."[25]

Many authors in the past have stated that the present space law regime does not prescribe a remedy for accidents and damages that result from GNSS signal defaults. Ingrid Lagarrigue[26] and Pablo Rodriguez-Contreras Perez[27] have ruled out the applicability of the Liability Convention of 1972[28] to the liabilities arising from CNS/ATM project. Ingrid Lagarrigue points out that "Neither the language of the 1972 convention, the negotiations leading to this convention and through established state practices could the 1972 convention be applied to indirect damages that result from the accidents of navigation satellite services provided by the states or an international organization."[29] "Rodriguez Perez also points out that neither the Russian Federation nor the United States (owners of the two operating systems) consider damage caused by the failure in GNSS signals to fall within the scope of the Liability Convention of 1972. Thereby, they do not accept international liability in such cases. But the author has suggested damage accruing to third countries from breach of duty to warn about error in signal would lead to the state considered "provider of signal" being held responsible under general customary international law."[30] It should also be taken into account that the drafters of Liability Convention would have never foreseen the applicability of the Liability Convention to the new genre of satellite based navigation services so it is quite doubtful whether Liability Convention would cover a damage that arises in the new context as discussed above. As far the applicability of provisions of the Outer Space Treaty of 1967 are concerned "in case damage results from satellite malfunction and consequent defective signaling it has been held that the primary provider of the signal could be held responsible if not liable under article 6 of this treaty."[31]

[25]Agenda Item: 2, LTEP/2, Second Meeting held on 6–10 October 1997, Paragraph 2:41 at pp. 2–8.

[26]Ingrid Lagarrigue, 'Are Existing Navigation satellite Liability Provisions adequate to govern navigation satellite malfunction, Outer Space Committee Newsletter, 2000.

[27]Pablo Rodriguez-Contreras Perez, 'GNSS Liability issues: Possible solutions to a global system', LLM Thesis, Institute of Air and Space Law, Mc gill University, 12th August 2002.

[28]This Convention will be referred to later in some detail elsewhere in this book.

[29]Ingrid Lagarrigue, 'Are Existing Navigation satellite Liability Provisions adequate to govern navigation satellite malfunction, Outer Space Committee Newsletter, 2000 at 206.

[30]Pablo Rodriguez-Contreras Perez, 'GNSS Liability issues: Possible solutions to a global system', LLM Thesis, Institute of Air and Space Law, Mc gill University, 12th August 2002 at 46.

[31]Pablo Rodriguez-Contreras Perez, 'GNSS Liability issues: Possible solutions to a global system', LLM Thesis, Institute of Air and Space Law, Mc gill University, 12th August 2002 at 45.

International liability could arise from hazardous or non hazardous activities of the state,[32] which involved a difference in the burden of proof to be discharged in either of the cases. The former involves no proof of existence of duty of care and violation of the said duty and later involves both the two previously mentioned elements to be proved. Thereby, now it becomes imperative to categorize "satellite malfunction" and "defective/faulty signal" as hazardous or non hazardous activity to be able to determine the levels of proof needed to be discharged. In case of malfunctioning of GNSS satellite run by US/Russian Federation malfunctions leading to defective signaling resulting in damage to a aircraft of third party be called a hazardous activity? "Hazardous nature of activity would depend on the risk that the activity entails and seriousness of the consequence arising from the failure of the activity."[33] In circumspection of two parameters mentioned above the signal errors and satellite malfunctioning leading to defective signaling may result in accidents of very grave consequences and is a very high risk prone activity, it should thereby be regarded as an hazardous activity entailing a strict international liability. "Even if international responsibility and liability were different concepts but one's who are responsible are usually the ones that were liable."[34]

The other aspects of the liability regime for the future is to effectively deal with the issue of sovereign immunities. As the requirement of Article 28 of the Chicago Convention of 1944 each Contracting State is responsible to provide air navigation facilities to facilitate international air navigation. Most of the state run air traffic providers enjoy immunity under federal statutes like for example the stature of Federal Tort Claims Act. In which case it is doubtful if a foreign plaintiff would be able to seek remedy from a State run GNSS facility. The LTEP Report of the second meeting admits that "most of the ATC agencies were run by the states, which were not willing to waive their sovereign immunity. The situation today is different. A new set of liability rules would be necessary to meet the new challenges and to give sufficient level of confidence to the new system."[35]It was also pointed out in the 1st LTEP meeting that "the principle of sovereign immunity was not applicable in cases of commercial transactions and further the predominant view was that it should be immaterial for purpose of liability if the service provider was a public or a private entity."[36]

Similarly, Henaku, in his paper entitled "Expanding Global Navigation Services: Selected Legal Issues" points outs that "the traditional disclaimers that were made available under the Telecommunications law should not be made available under GNSS navigation services as he believes "In addition to the technical services, navigation satellites would be playing a more important role in assurance of safety

---

[32]Ingrid Lagarrigue, 'Are Existing Navigation satellite Liability Provisions adequate to govern navigation satellite malfunction, Outer Space Committee Newsletter, 2000 at 36.

[33]Agenda Item: 2, LTEP/2, Second Meeting held on 6–10 October 1997, Paragraph 2:41 at 39.

[34]ICAO, LTEP/1, 25–30th November 1996, Montreal, Paragraph 3:32 at 3–5.

[35]Agenda Item: 2, LTEP/2, Second Meeting held on 6–10 October 1997, Paragraph 2:40 at 2–8.

[36]ICAO, LTEP/1, 25–30th November 1996, Montreal, Paragraph 3:37 at 3–5.

of life. While human loss is imaginable as a result of a telecommunication breakdown, the likelihood, extent and impact would be pale compared with a navigation satellite failure."[37]

LTEP in its first meeting had suggested the concept of channeling of liability as it was realized that "navigation through GNSS involved many chains of providers and users" so a definite scheme of liability allocation needed to be worked upon. So channeling involved "establishment of a rule allocating ultimate liability to an appropriate place in chain of providers and users. In the Working paper presented by EUROCONTROL during the aforesaid meeting a regulatory chain was also proposed, including users, states, infrastructure organization and system operators."[38]

References

Henaku BDK (1999) Expanding global navigation services: selected legal issues. In: Proceedings of workshop of space law on space law in the 21st century, p 174

Huang J (1997) Development of the long term legal framework for global navigation satellite system. Annals of Air Space Law 22-1:594

Kolodziej K (2006) Location 2006 conference – a report from India. Directions Magazine. Information available from http://www.directionsmag.com/article.php?article_id=2232&trv=1. Accessed on 23 Aug 2006

Lammerstma PF (2005) Satellite navigation. Institute of Information and Computing Science, Utrecht University, Utrecht, p 14

Lawson R (2004) The Space Security Index. Astropolitics. Int J Space Polit Policy 2(2):177, 1557–2943

[37]Henaku (1999), p. 174.

[38]Huang (1997), p. 594.

Chapter 3
Is There Life in Outer Space?

In a thought provoking commentary in the *Journal of Space Law*, Marcia S. Smith envisions that if life were to be found in outer space the ensuing conduct of human kind would be more an ethical issue.[1] Any inquiry into life in outer space has to take into account both vegetation as well as intelligent life. In both instances the main consideration would be how we could protect such life forms and use them for the benefit of human kind. The central theme of space exploration would incontrovertibly be international cooperation and abstinence from the use of force, which collectively form the cornerstone of space exploration from a legal standpoint.

Ms. Smith asks the pertinent questions: "Do we send more probes to further investigate and do we have a responsibility to protect that life and allow it to develop naturally? If robotic probes definitively find life, should we erect a 'do not disturb sign' and refrain from sending further probes?" While these questions would have to be asked and answered at one point or another, the more immediate issue would be what we would do on Earth to cope with the new exigency.

One of the corollaries to finding life in outer space would be the issue of how we would use such a discovery in the context of the prevailing environment of international relations. In this context international politics within the umbrella of the United Nations and the United Nations Charter may become extremely relevant. It is not unrealistic to envision that the discovery of life in outer space could spark a discourse on interests and a renewed initiative to revisit international treaties to ensure the peaceful uses of outer space while at the same time ensuring some degree of control on the use of life so discovered.

At the 79th Plenary Meeting of its Sixty first Session, the United Nations General Assembly adopted Resolution 61/111 which, *inter alia*, expresses serious concern of the General Assembly about the possibility of an arms race in outer space and urges all States, in particular those with major space capabilities, to contribute actively to the goal of preventing an arms race in outer space as an essential condition for the promotion of international cooperation in the exploration and use of outer space for peaceful purposes. Doubtless, such a threat would prove to be more real and ominous if life in outer space were to be discovered.

---

[1] Smith (2006), pp. 217–222.

R. Abeyratne, *Space Security Law*,
DOI 10.1007/978-3-642-16702-7_3, © Springer-Verlag Berlin Heidelberg 2011

The General Assembly also agreed that a panel of space exploration activities, including the participation of the private sector should be convened during the fiftieth session of the United Nations Committee for the Peaceful Uses of Outer Space (UNCUPUOS). Perhaps the most noteworthy of the Assembly's observations as recorded in the Resolution is that the recommendations of UNISPACE III could be integrated into the work programme of the Office of Outer Space Affairs and that UNCUPUOS could consider these recommendations for implementation. UNISPACE III is the genesis of the Vienna Declaration which, *inter alia* espouses the protection of the outer space environment.

Although the title of this article is purely conjectural, its contents – in the eventuality of the occurrence suggested therein becoming real – bring to bear the need to reflect on principles of human conduct and liability prescribed by existing legal norms and ethical considerations with regard to the treatment of life. American geneticist Joshua Lederberg introduced to the world the science of exobiology (or astrobiology) – a branch of biology which deals with the search for extraterrestrial life, especially intelligent life, outside the solar system. Although remote astronomical observations of a planet or other celestial body provide information about its physical environment, the determination of the presence of life on these bodies is more difficult. Exobiological techniques are designed to detect life forms, artefacts produced by intelligent life, waste produces of metabolic reactions, remnants of former life, pre-biological molecules that may reflect early evolutionary stages or substances such as carbon or combination of Hydrogen and Oxygen forming water that are necessary for the sustenance of life as it is experienced on Earth.

In 1948 the U.S. Air Force commenced maintaining a file of reports relating to extraterrestrial phenomena called Project Blue Book. In July 1952, the U.S. government established a panel of scientists including engineers, meteorologists, physicists and an astronomer to investigate a series of radar detection coincident with visual sightings near the national airport in Washington D.C. The panel was organized by the Central Intelligence Agency, which underscores the thrust of public and government concern and interest at the time.

The concern was based on U.S. military activities and intelligence and that its report was originally classified Secret. Later declassified, the report revealed that 90 per cent of UFO sightings could be readily identified with astronomical and meteorological phenomena (*e.g.* bright planets, meteors, auroras, ion clouds) or with aircraft, birds, balloons, searchlights, hot gases, and other phenomena, sometimes complicated by unusual meteorological conditions.

The publicity given to early sightings in the press undoubtedly helped stimulate further sightings not only in the U.S. but also in Western Europe, the Soviet Union, Australia, and elsewhere. A second panel established in February 1966 reached conclusions similar to those of its predecessor. This left a number of sightings admittedly unexplained, and in the mid-1960s a few scientists and engineers, notably James E. McDonald, a University of Arizona meteorologist, and J. Allen Hynek, a Northwestern University astronomer, concluded that a small percentage of the most reliable UFO reports gave definite indications of the presence of extraterrestrial visitors.

This sensational hypothesis, promoted in newspaper and magazine articles, met with prompt resistance from other scientists. The continuing controversy led in 1968 to the sponsorship by the U.S. Air Force of a study at the University of Colorado under the direction of E.U. Condon, a noted physicist. The Condon Report, "A Scientific Study of UFO's" was reviewed by a special committee of the National Academy of Sciences and released in early 1969. A total of 37 scientists wrote chapters or parts of chapters for the report, which covered investigations of 59 UFO sightings in detail, analyzed public-opinion polls and reviewed the capabilities of radar and photography. Condon's own "Conclusions and Recommendations" firmly rejected ETH – the extraterrestrial hypothesis – and declared that no further investigation was needed.

This left a wide variety of opinions on UFO's. A large fraction of the U.S. public, and a few scientists and engineers, continued to support ETH. A middle group of scientists felt that the possibility of extraterrestrial visitation, however slight, justified continued investigation, and still another group favoured continuing investigation on the grounds that UFO reports are useful in socio-psychological studies. These varying views and attitudes were expressed at a symposium held by the American Association for the Advancement of Science, in December 1969. Several years later, in 1973, a group of U.S. scientists organized the Centre for UFO Studies in Northfield, Ill., to conduct further work.

On 24 June 1947, American pilot Kenneth Arnold described some unusual flying craft he had seen over the mountains off the west coast of the United States. In his words: "they flew like a saucer would if you skipped it across water." Newspapers applied his phrase to the craft themselves, and the misleading label "flying saucer" has followed the phenomenon of the unidentified flying object (UFO) since.

The Word UFO – unidentified flying object – officially means simply something that has not been, or cannot be accounted for by any of the known laws of physics. But the seemingly rational behaviour reported in many UFO sightings, as well as the accounts of meetings with humanoids, has led to the speculation that UFOs are, in fact, spacecraft bringing creatures from outer space.

If this is so, the spaceships must be able to cover immense distances. People who claim to have had contact with extra-terrestrials often say they have spoken with Venusians. But Venus is highly unlikely to be inhabited. Any intelligent life forms must be coming from still further away, and, even assuming that life spans of creatures from other planets may be much longer than our own, it is clear that UFOs must be able to travel very fast indeed if they are not to take hundreds of years to travel between inhabited planets.

Reports of the movement of UFOs are remarkably consistent. Most people describe them as hovering and then taking off at very high speed, often executing manoeuvres that would be impossible in conventional aircraft. Even allowing for exaggeration by excited witnesses, the consistency of the reports suggests the UFOs use a very powerful force to produce dramatic accelerations.

None of the rocket fuels we use at present can produce either the speed or acceleration observed in the UFOs, because they store only a small amount of energy for a given mass. Right from the beginning, rocket travel has been faced with the problem of enabling the rocket to carry enough fuel for its journey – it must lift the fuel, which can be very heavy if the journey is long, as well as itself and its occupants. The solution has been the multi-stage rocket: the initial acceleration is given by a rocket that is jettisoned when its fuel is used up and a second rocket takes over.

Space flights have always stretched our rocket technology to its limits – and, as everyone knows, our rockets and spacecraft do not accelerate very briskly away from the Earth. Although they eventually reach quite high speeds, they are nowhere near fast enough to reach planets outside our Solar System within a human lifetime.

If we assume UFOs are subject to the same laws of physics as we are, then, to operate on and near the Earth with the rapid accelerations and manoeuvres at high speeds that are often reported, they must be using a different source of energy from conventional chemical fuels. Their fuel must be highly compact, with a high energy yield for a small mass: the obvious source is nuclear fuel.

As long ago as 1958 – just after Yuri Gagarin had become the first man ever to orbit the Earth in space – Freeman Dyson, a theoretical physicist, embarked on a plan for a nuclear-powered spaceship. He had previously worked on the development of the atom bomb and had a comprehensive understanding of nuclear power. He assembled a group of scientists at La Jolla, Southern California, to work with him; he called his scheme "Project Orion."

Project Orion was a serious attempt to build a spacecraft powered by nuclear explosions, and was intended as an alternative to the multi-stage rockets that Werner von Braun was proposing for space travel.

Freeman Dyson's ultimate aim was to build a spacecraft the size of a small city that would take a group of people to a distant comet on the edge of the solar system, where they would settle. This may have been only a pipe-dream, but the design was real enough.

The spacecraft was to be powered by hydrogen bombs. Essentially, his idea was to carry a number of hydrogen bombs aboard the spacecraft; these would be moved, one by one, to a position underneath the craft where they would be exploded. The base of the spacecraft would absorb the shock and the craft would be driven along. Obviously the spacecraft and the bomb system would have to be designed so that the craft was propelled along and not simply blown apart, but – in principle, at least – this was straightforward. However, Dyson was never able to test his ideas: he was prevented by public concern about the pollution of the atmosphere by radioactive fallout.

UFOs are often reported as disappearing rapidly – going of "like a television set" and reappearing just as quickly. This aspect of the phenomenon has puzzled scientists for a long time and has led to suggestions that UFOs use some kind of "anti-optic device" to prevent them from being seen. There are, however, some simpler explanations that account for the majority of reports. UFOs "disappearing" in the of night could do so by simply switching off their lights' daytime discs could appear to vanish by turning themselves sideways on to the observer – it would be very difficult to pick out the thin edge of a disc against the sky. These explanations do not, of course, account for radar-visual sightings that suddenly vanish. But if a UFO disappeared behind a patch of disturbed air, a mirage-like effect could easily screen it both from sight and from radar detectors.

There are, however, cases on file for which none of these explanations seem credible. It seems that the phenomena involved can only be explained as products of a technology much further advanced than our own.

By far the majority of UFO reports describe the strange objects as disc- or cigar-shaped and it could be that most UFOs reported as cigar-shaped are in fact discs. Whether or not this is actually the case, the number of reports of saucer-shaped UFOs is overwhelming. There has been a great deal of speculation as to why this should be so – some people have suggested the mystical significance of the circle may have something to do with it – but there is a simple explanation.

On long inter-stellar voyages, a spacecraft will pass through vast regions of empty space – far from the regions of gravitational attraction of any major objects – where there is no wind resistance, no up or down, no east or west, nothing. The most logical shape for a vessel travelling in these circumstances is circular, for a circle is symmetrical about a infinite number of axes. The fact that most UFOs are disc-shaped rather than spherical can be explained as a design feature that allows spacecraft to operate at high speeds once they have entered the atmosphere of planets: by flying with their edges into the wind, they can cut down the effect of air resistance almost to zero.

Whether or not UFOs existed in the past, there is no doubt that UFO sightings have proliferated in astonishing numbers over the past 30 years. This fact seems to be in some way linked with man's first steps towards exploring space, and this connection is undoubtedly an important clue in trying to explain the UFO.

Estimates of the total number of UFO sightings vary so widely as to be meaningless; more helpful figures are provided by the catalogues of reported sightings prepared by individual investigative organizations. In the 1960s a French team catalogued more than 600 encounter cases in France alone, each vouched for by responsible investigators. In the early 1970s UFO investigators made lists of all reported landing cases for particular countries: 923 were recorded in the United States, 200 in Spain.

Are UFOs real in the sense that, say, spacecraft are real? The surest proof would be actually to get hold of one, and there are persistent rumours that certain governments, notably that of the United States, have indeed obtained a UFO, which is kept in total secrecy. However this remains mere conjecture, despite the sworn affidavits of alleged witnesses. Indeed, the whole matter of governmental involvement – or the lack of it – is a further and fascinating aspect of the UFO controversy.

In the absence of a real UFO that we can touch and examine, there is a great deal of evidence of the phenomenon in the form of a mass of photographs and a handful of movies. The majority are undoubtedly fakes. Those with good credentials are so blurred, so distant or so ambiguous that they simply add a further dimension to the problem: why, if UFOs exist, and in an age when many people carry cameras with them most of the time, have we not obtained better photographic evidence?

Perhaps the strongest evidence we have is from the effects caused by UFOs on surrounding objects, particularly machinery. In November 1967 a truck and a car approaching each other on a Hampshire road in the United States in the early hours of the morning simultaneously suffered engine failure when a large egg-shaped object crossed the road between them. The police, and subsequently the Ministry of Defence, investigated the incident, but no official explanation was ever issued. Such a case may leave investigators puzzled, but it makes one thing certain: if they can cause physical effects, UFOs must be physically real.

If they are physical objects, UFOs must originate from somewhere. When the first UFOs of the current era were seen, back in the 1940s, it was assumed they came from somewhere on Earth. The Americans suspected they were a Russian secret device, perhaps developed using the expertise of German scientists captured at the end of the Second World War.

But as more reports came in it became clear that no nation on Earth could be responsible. Nor was there sufficient evidence to support other ingenious theories – that they came from the Himalayas, long a favoured source of secret wisdom, or Antarctica, where unexplored tracts of land and climatic anomalies provide a shaky foundation for speculation. Instead, ufologists began to look beyond the Earth, encouraged by the fact that our own space explorations programme was just beginning. We were starting to take an active interest in worlds beyond, and it seemed reasonable that other civilizations might have a similar interest in us.

However, although the number of potential sources of life in the Universe is virtually infinite, the probability of any civilisation being at a stage of development appropriate for space travel is very small. The fact that no solid evidence has been found for the extraterrestrial hypothesis is discouraging. Although it is the best available explanation, it remains no more than speculation.

The *Moon Agreement* of 1979, which will be discussed later in this book, provides that in the exploration and use of the moon, States Parties shall take measures *inter alia* to avoid harmfully affecting the environment of the earth through the introduction of extra terrestrial matter or otherwise. At the same time, the Moon agreement, which applies to other celestial bodies as well, in Article 6 provides that there shall be freedom of scientific investigation on the moon by all States Parties without discrimination of any kind, on the basis of equality and in accordance with international law. The same provision allows a State Party which finds minerals and other substances to collect and take custody of such material and even conduct scientific experiments on the samples. They are required to have regard to requests from other State Parties for the use of such material and also given the discretion to exchange scientific information in this context. In this context it must be noted that at the frontiers of this issue are the astronauts, who are by treaty designated as envoys of mankind in outer space, casting on them the responsibility of adhering to applicable treaty provisions on behalf of their States.

In the field of international space law, two clearly connected terms have been used: liability and responsibility. Although "responsibility" has not been cohesively interpreted in any legal treaty relating to outer space, "liability" occurs in the *Convention on International Liability for Damage Caused by Space Objects*, March 29 1972 (Liability Convention) and is sufficiently clear therein. This, however, does not mean that State responsibility is not relevant to the obligations of States law as, in international relations, the invasion of a right or other legal interest of one subject of the law by another inevitably creates legal responsibility.

At present, one can regard responsibility as a general principle of international law, a concomitant of substantive rules and of the supposition that acts and omissions may be categorized as illegal by reference to the rules establishing rights and duties. Shortly, the law of responsibility is concerned with the incidence and consequence of illegal acts, and particularly the payment of compensation for loss caused. Therefore, As discussed, both treaty law and general principles of international law on the subject of space law make the two elements of liability and responsibility a means to an end – that of awarding compensation to an aggrieved State or other subject under the law. In view of the many legal issues that may arise, the primary purpose of a regulatory body which sets standards on State liability would be to carefully consider the subtleties of responsibility and liability and explore their consequences on States and others involved as they apply to the overall concept of the status of a State as a user of space technology which may cause harm or injury to the latter.

Finally, we have to be mindful of a few fundamental truths. First, if we come across any form of life in outer space it will be the concern of all humankind. Second, any treatment of such life, irrespective of the fact that it is found in outer space, should be according to the principles of international law and the United Nations Charter, which contains numerous provisions which are relevant to the use of force. Several General Assembly resolutions, adopted without dissent or with near unanimity, have restated, amplified and clarified the meaning of these Charter provisions. Possibly, the most relevant and authoritative is the 1970 Declaration on Principles of International Law Concerning Friendly Relations and Co-operation Among States in Accordance with the Charter of the United Nations (Friendly Relations Declaration). Although some dispute exists in regard to the precise legal status of the *Friendly Relations Declaration*, it is generally regarded as an authoritative interpretation of broad principles of international law expressed in the Charter. Another, and more controversial example, is the General Assembly's "Definition of Aggression" resolution.

The UN Charter does not directly address the question of intervention by states; rather, under Article 2(7) it precludes the Organization itself from intervening "in matters which are essentially within the domestic jurisdiction of any state." Hence the General Assembly's Declaration on the Inadmissibility of Intervention in the Domestic Affairs of States and the Protection of Their Independence and Sovereignty, adopted in 1965 by a vote of 109 to none with one abstention, takes on special legal significance.

Within these parameters, yes, we could send more probes to investigate further. Yes, we could even put up a "do not disturb sign." But whatever we do, we are bound by the principles of responsibility and international accountability to treat life in outer space with the same dignity that should be accorded to life on Earth and to strictly adhere to the principles of international law and the United Nations Charter in ensuring peace in the process.

What Should We Do If We Find Life in Outer Space?

Humanity's foray into the solar system brings out the ethical issue of what we should do if life is found in outer space. Do we send more probes to further investigate and do we have a responsibility to protect that life and allow it to develop naturally? If robotic probes definitively find life, should we erect a "do not disturb sign" and refrain from sending further probes? Then again, what if we were to find intelligent life forms closer to the human form and not mere vegetation? Could we exercise control over the welfare of such life and who would claim that control?

Space law is grounded on the principle that outer space is the common heritage of mankind and that no State or individual can therefore claim *rights in rem* to any portion of outer space. Air law, on the other hand, is firmly entrenched in the principle of sovereignty of States, so that a State may lay claims to rights over the airspace above its territory. This essentially means that while the implementation of air law is heavily influenced by municipal law, space law is solely grounded on legal principles binding on the community of nations. Principles of public international law therefore play an exclusive part in the application of space law principles.

Extraterrestrial life is life originating outside of the Earth. Its existence remains hypothetical; there is no evidence of extraterrestrial life that has been widely accepted by the scientific community. Most scientists believe that if extraterrestrial life exists, its emergence occurred independently, in different places in the universe. An alternative hypothesis is panspermia, which suggests that life might emerge in one location and then spread between habitable planets. These two hypotheses are not mutually exclusive. The study and theorization of extraterrestrial life is known as astrobiology, exobiology or xenobiology.

American geneticist Joshua Lederberg introduced to the world the science of exobiology (or astrobiology) – a branch of biology which deals with the search for extraterrestrial life, especially intelligent life, outside the solar system. Although remote astronomical observations of a planet or other celestial body provide information about its physical environment, the determination of the presence of life on these bodies is more difficult. Exobiological techniques are designed to detect life forms, artifacts produced by intelligent life, waste produces of metabolic reactions, remnants of former life, prebiological molecules that may reflect early evolutionary stages or substances such as carbon or combination of Hydrogen and Oxygen forming water that are necessary for the sustenance of life as it is experienced on Earth.

In 1948 the U.S. Air Force commenced maintaining a file of reports relating to extraterrestrial phenomena called Project Blue Book. In July 1952, the U.S. government established a panel of scientists including engineers, meteorologists, physicists and an astronomer to investigate a series of radar detection coincident with visual sightings near the national airport in Washington D.C. The panel was organized by the Central Intelligence Agency, which underscores the thrust of public and government concern and interest at the time.

The concern was based on U.S. military activities and intelligence and that its report was originally classified Secret. Later declassified, the report revealed that 90 per cent of UFO sightings could be readily identified with astronomical and meteorological phenomena (*e.g.* bright planets, meteors, auroras, ion clouds) or with aircraft, birds, balloons, searchlights, hot gases, and other phenomena, sometimes complicated by unusual meteorological conditions.

The publicity given to early sightings in the press undoubtedly helped stimulate further sightings not only in the U.S. but also in Western Europe, the Soviet Union, Australia, and elsewhere. A second panel established in February 1966 reached conclusions similar to those of its predecessor. This left a number of sightings admittedly unexplained, and in the mid-1960s a few scientists and engineers, notably James E. McDonald, a University of Arizona meteorologist, and J. Allen Hynek, a Northwestern University astronomer, concluded that a small percentage of the most reliable UFO reports gave definite indications of the presence of extraterrestrial visitors.

This sensational hypothesis, promoted in newspaper and magazine articles, met with prompt resistance from other scientists. The continuing controversy led in 1968 to the sponsorship by the U.S. Air Force of a study at the University of Colorado under the direction of E.U. Condon, a noted physicist. The Condon Report, "A Scientific Study of UFO's" was reviewed by a special committee of the National Academy of Sciences and released in early 1969. A total of 37 scientists wrote chapters or parts of chapters for the report, which covered investigations of 59 UFO sightings in detail, analyzed public-opinion polls and reviewed the capabilities of radar and photography. Condon's own "Conclusions and Recommendations" firmly rejected ETH – the extraterrestrial hypothesis – and declared that no further investigation was needed.

This left a wide variety of opinions on UFO's. A large fraction of the U.S. public, and a few scientists and engineers, continued to support ETH. A middle group of scientists felt that the possibility of extraterrestrial visitation, however slight, justified continued investigation, and still another group favoured continuing investigation on the grounds that UFO reports are useful in sociopsychological studies. These varying views and attitudes were expressed at a symposium held by the American Association for the Advancement of Science, in December 1969. Several years later, in 1973, a group of U.S. scientists organized the Center for UFO Studies in Northfield, Ill., to conduct further work.

SETI (Search for Extra-Terrestrial Intelligence) is the collective name for a number of organized efforts to detect extra terrestrial life. A number of efforts with SETI in the project name have been organized, including projects funded by the United States Government. The generic approach of SETI projects is to survey the sky to detect the existence of transmissions from a civilization on a distant planet, an approach widely endorsed by the scientific community as hard science.

"What are we entitled to do if we found extra terrestrial life?" Firstly, the *Outer Space Treaty* of 1967 in Article II provides that outer space, including the Moon and other celestial bodies is not subject to national appropriation by claim or sovereignty by means of use or occupation, or by any other means. Furthermore the Treaty, in Article III requires that States parties to the Treaty shall carry on activities in the exploration and use of outer space, including the Moon and other celestial bodies, according to the principles of international law, including the Charter of the United Nations, in the interest of maintaining international peace and security and promoting international cooperation and understanding.

There is no known principle or pronouncement of law which mentions "extra-terrestrial intelligence." Therefore, no existing norm of international law prohibits social intercourse in outer space between humans and extraterrestrial beings (if such were to exist). The only prohibition which will obtain at international law would be if such intercourse leads to the "non peaceful" use of outer space by States. It is incontrovertible that the absence of peaceful use of outer space (when states indulge in activities in outer space) would inevitably mean warlike or aggressive use of outer space. Accordingly, such action would perforce form "use of force" by such states on other states or persons affected by these actions.

In a purely forensic sense, the fundamental postulate of space law, which devolves upon States the responsibility to explore and use outer space for peaceful purposes is pivotal to the conjecture of extra terrestrial intelligence. To start with there were two United Nations General Assembly resolutions on the subject. Resolution 49/34 of 1994, which *inter alia* covered international co-operation in the peaceful uses of outer space, links the importance of international co-operation in developing the rule of law, including the relevant norms of space law and their important role in international co-operation for the exploration and use of outer space for peaceful purposes.

The operative criterion for the adoption of the above aims, as used by the Resolution, lies in the endorsement that they should be achieved through international co-operation in the development of the role of law. The Resolution therefore brings to bear the ineluctable and compelling need for the application of existing principles of international law as a means towards this end.

The General Assembly of the United Nations, on 5 February 1996, adopted Resolution 56/27 relating to international co-operation in the peaceful uses of outer space. This resolution broadly reaffirmed the principles of Resolution 49/34. Having considered the Report of UNCOPUOS on the work of its thirty-eighth session, the General Assembly endorsed the Committee's recommendation that the Committee should, through its Scientific and Technical Sub-Committee, *inter alia*, consider the use of nuclear power sources in outer space and questions relating to space transportation systems and their implications for future activities in space. Matters relating to life sciences, space medicine and astronomy were some of the areas that were focused on for further consideration in the Resolution.

On 14 December 2006, at its 79th Plenary Session, The United Nations General Assembly adopted Resolution 61/111 on international cooperation in the peaceful uses of outer space. This Resolution requests *inter alia* UNCOPUOS to consider, as a matter of priority ways and means of maintaining outer space for peaceful purposes.

Article VI of the *Outer Space Treaty* provides in part that State Parties to the Treaty shall bear international responsibility for national activities in outer space, whether such activities are carried out by governmental agencies or non-governmental agencies. This provision clearly introduces the notion of strict liability *erga omnes* to the application of the *jus cogens* principle relating to outer space activities of States and could be considered applicable in instances where States hold out to the international community as providers of technology achieved and used by them in outer space, which is used for purposes of air navigation. Article VI further requires that the activities of non-governmental entities in outer space shall require authorization and continuing supervision by the appropriate State Party to the Treaty, thus ensuring that the State whose nationality the entity bears would be vicariously answerable for the activities of that organization, thereby imputing liability to the State concerned.

The *Moon Agreement* of 1979 provides that in the exploration and use of the moon, States Parties shall take measures *inter alia* to avoid harmfully affecting the environment of the earth through the introduction of extra terrestrial matter or otherwise. In the field of international space law, two clearly connected terms have been used: liability and responsibility. Although "responsibility" has not been cohesively interpreted in any legal treaty relating to outer space, "liability" occurs in the Convention on International Liability for Damage Caused by Space Objects, March 29 1972 (Liability Convention) and is sufficiently clear therein. This, however, does not mean that State responsibility is not relevant to the obligations of States law as, in international relations, the invasion of a right or other legal interest of one subject of the law by another inevitably creates legal responsibility.

At present, one can regard responsibility as a general principle of international law, a concomitant of substantive rules and of the supposition that acts and omissions may be categorized as illegal by reference to the rules establishing rights and duties. Shortly, the law of responsibility is concerned with the incidence and consequence of illegal acts, and particularly the payment of compensation for loss caused. Therefore, As discussed, both treaty law and general principles of international law on the subject of space law make the two elements of liability and responsibility a means to an end – that of awarding compensation to an aggrieved State or other subject under the law. In view of the many legal issues that may arise, the primary purpose of a regulatory body which sets standards on State liability would be to carefully consider the subtleties of responsibility and liability and explore their consequences on States and others involved as they apply to the overall concept of the status of a State as a user of space technology which may cause harm or injury to the latter.

Finally, we have to be mindful of a few fundamental truths. Firstly, if we come across any form of life in outer space it will be the concern of all humankind.

Second, any treatment of such life, irrespective of the fact that it is found in outer space, should be according to the principles of international law and the United Nations Charter. Within these parameters, yes, we could send more probes to investigate further. Yes, we could even put up a "do not disturb sign." But whatever we do, we are bound by the principles of responsibility and international accountability to treat life in outer space with the same dignity accorded to life on Earth.

Reference

Smith MS (2006) The vision for space exploration: expanding the envelope for space law debates. J Space Law 32(1):217–222

Chapter 4
Space Tourism

Narrowly defined, the word "tourism" means travel for recreation or instruction, often in organized groups. The tourism industry primarily provides the tourist with travel to the destination and thereafter provides accommodation usually in a commercial establishment that provides lodging, food, and other services to the public. Therefore tourism is essentially associated with the transport and hospitality industries, where the hotel business features as an important industry which caters to people traveling for business or pleasure. When these factors are translated into exigencies of a viable space tourism industry, many considerations emerge, particularly from an extra terrestrial perspective. The main issues are whether a commercially viable and sufficiently evolved space transportation program could be a reality in the near future and whether the infrastructure needed for establishing accommodation for a sustained tourism industry in the inhospitable terrain of outer space could be put into place. Some have suggested that space tourism is indeed a realistic goal in the near future particularly if a space program were calculated to create permanent settlements. The residents of such outposts would have to "live off the land," obtaining necessities such as oxygen and water from the harsh environment of outer space. For example, it has been suggested that on the Moon, pioneers could obtain oxygen by heating lunar soil. In 1998 the Lunar Prospector discovered evidence of significant deposits of ice – a valuable resource for settlers – mixed with soil at the lunar poles. It is also known that on Mars, oxygen could be extracted from the atmosphere and water could come from buried deposits of ice.

Space tourism, which would have been merely a dream and a cinematographic fantasy at best is fast becoming neither a fantasy nor just a nickname for conventional manned space flights. It is now considered a viable economic activity based on public demand. Dennis Tito and Mark Shuttleworth, both of whom traveled as tourists in space have already obviated any doubts regarding the immense possibilities of this activity. Space tourism is a term broadly applied to the concept of travel beyond Earth's atmosphere by paying customers. It can be defined to include not only the vehicles that take public passengers into space, but also from the perspective of the "destination" paradigm. As such, the industry can be envisioned to include earth-based attractions that simulate the space experience such as space theme parks, space training camps, virtual reality facilities, multi-media interactive

R. Abeyratne, *Space Security Law*,
DOI 10.1007/978-3-642-16702-7_4, © Springer-Verlag Berlin Heidelberg 2011

games, and telerobotic moon rovers controlled from earth. Also included are parabolic flight, vertical suborbital flights, orbital flights lasting up to 3 days, or week-long stays at a floating space hotel, including participatory educational, research and entertainment experiences as well as space sports competitions (i.e., space Olympics).

To see the unseen and know the unknown has been the genesis and heritage of human aspiration from early times, resulting in human migration and travel over centuries. The arcane desire to conquer the invincible is an endemic human trait. Space tourism has the added dimension of making space tourists ambassadors of planet Earth to other celestial territories while at the same time giving them the thrill of crossing the frontiers of the Earth's atmosphere into uncharted territory that is outer space. It is believed that the sensation of weightlessness and the defeat of the force of gravity are the most alluring to the space tourist. Recent advancements in space technology have enabled the world community to develop safe, reliable and affordable transportation systems for space travel within the next decade or so. The National Aerospace Laboratory of Japan, in a recent market survey on space tourism, has revealed that the price of a return ticket to low Earth orbit should be reduced to between US $ 10,000 to US$ 20,000 per person. A market of one million passengers per year from the world's two largest markets – North America and Europe – would, at US$ 10,000 per return ticket, yield revenues of US$ 10 billion a year. This would make space travel by the ordinary or "average" citizen of the world a common occurrence. It is reported that Enzo Paci, Chief Statistician of the World Tourism Organization, has conducted a recent study in which he concludes that short pleasure voyages to outer space by tourists will become a reality even as early as 2004 or 2005.[1]

Taken from a socio-legal perspective, space tourism brings to bear unique considerations, from the status of the space tourist to the conduct expected of such a person and the various liability regimes that might be required to address the "package deal" concerning the contract of carriage to outer space and amenities provided by the service provider. Additionally, real concerns of liability, insurance coverage and risk management would have to be allayed before a sustained space tourism programme takes to the heavens. Analogically, when a serious commercial air transport regime took off after World War II, a dedicated international organization – the International Civil Aviation Organization (ICAO) – was established through international treaty. The purpose of ICAO was, and continues to be to implement the various provisions of the Convention and its 18 Annexes containing rules and regulations which comprehensively cover all perceivable aspects of international civil aviation.[2] The question which now arises is whether commercial space travel will require such an organization driven by an international treaty. It is also pertinent to determine the extent to which the aviation process will apply to a space travel regime.

---

[1]Frangialli (1999), p. 2.

[2]Schenkman (1955), pp. 120–121.

Who Is a Space Tourist?

In Space Law, there is no such being as a "person" in outer space. There are only astronauts and personnel.[3] The 1967 Outer Space Treaty stipulates that State parties to the Treaty shall regard astronauts as envoys of mankind in outer space and shall render to them all possible assistance in the event of accident, distress or emergency landing on the territory of another State party or on the high seas.[4] The provision also requires State parties to return astronauts under the above circumstances safely and promptly to the State of registry of their space vehicle.[5]

The Treaty provision is a reproduction *verbatim* of Paragraph 9 of United Nations General Assembly Resolution XVIII of 1962.[6] Although initially, the world's "envoys of mankind" seemingly created some apprehension in the international community as to whether such phraseology connoted diplomatic immunity to astronauts, Bin Cheng clears up this ambivalence by concluding that it was only a figure of speech which has not been repeated in any United Nation's documentation yet.[7] The perceived inadequacy of definitive identification at international law of an astronaut and his conduct in outer space leaves one with the basic premise that international law is incontrovertibly applicable to outer space activities and outer space, including the moon and other celestial bodies, which are totally independent of appropriation by States or individuals. This in turn leaves one with the inevitable conclusion that outer space would be analogous at international law to the high seas.

Jurisdictionally, any person comes clearly within the purview of the State on whose territory he is or above whose territory and in the airspace of the State concerned, if he is in an aircraft. Generally, in outer space, this *status quo* changes,

---

[3]Treaty on Principles Governing the Activities of States in the Exploration and Use of Outer Space, Including the Moon and Other Celestial Bodies, Opened for Signature at Moscow, London and Washington on 27 January 1967, 610VNTS 205. It must be noted that the first "space tourist" Denis Tito was called a space tourist for purposes of public reference by the media. He was called a "guest cosmonaut" by the Russians and an amateur astronaut by the Americans. See http://www.spacedaily.com/news/011206133411.3i4zwq28.html.

[4]Treaty on Principles Governing the Activities of States in the Exploration and Use of Outer Space, Including the Moon and Other Celestial Bodies, Opened for Signature at Moscow, London and Washington on 27 January 1967, 610VNTS 205. It must be noted that the first "space tourist" Denis Tito was called a space tourist for purposes of public reference by the media. He was called a "guest cosmonaut" by the Russians and an amateur astronaut by the Americans. See http://www.spacedaily.com/news/011206133411.3i4zwq28.html. Article V.

[5]Treaty on Principles Governing the Activities of States in the Exploration and Use of Outer Space, Including the Moon and Other Celestial Bodies, Opened for Signature at Moscow, London and Washington on 27 January 1967, 610VNTS 205. It must be noted that the first "space tourist" Denis Tito was called a space tourist for purposes of public reference by the media. He was called a "guest cosmonaut" by the Russians and an amateur astronaut by the Americans. See http://www.spacedaily.com/news/011206133411.3i4zwq28.html. Article V.

[6]UNGA Resolution 1962 (XVIII) Declaration of Legal Principles Governing the Activities of States in the Exploration and Use of Outer Space.

[7]Cheng (1997), pp. 259–460.

and astronauts become liable under the laws of the State of registry or the State which launches their spacecraft for travel and work in outer space.[8] This is brought to bear by Article VIII of the Outer Space Treaty which provides:

> A State party to the Treaty on whose registry an object launched into outer space is carried shall retain jurisdiction and control over such object, and over any personnel thereof, while in outer space or on a celestial body.[9]

However, as Bin Cheng validly points out[10] the interpretation of Article VIII could well result in ambivalence and confusion. The "object" and "personnel" referred to in the Treaty provision do not adequately cover persons who are not "personnel" such as passengers in a spacecraft. Of course, as Cheng maintains, the quasi jurisdiction of the State of registry of the spacecraft can apply both in the instance of conduct in the spacecraft as well as outside the spacecraft on the basis that the astronaut concerned would be deemed to belong to the spacecraft at all times in outer space. Logically, therefore, such jurisdiction could be imputed to passengers, visitors and guests by linking them to the spacecraft in which they travelled. This far reaching generalization would then cover the conduct of an astronaut or other persons while walking on the moon, Mars or other celestial body, as well as such persons who go on space walks outside the spacecraft in which they travelled.

Another provision which sheds some light on past attempts by the international community to identify liability and jurisdictional issues relating to astronauts is Article 12 of the Moon Treaty of 1979[11] which provides:

> States Parties shall retain jurisdiction and control over their personnel, space vehicles, equipment facilities, stations and installations on the moon...

It is presumed that the legal link between the personnel and the spacecraft they travel in under the circumstances are imputed to the State of registry of the said craft. If this were not the case, and such a link cannot be established, the provision itself becomes meaningless and destitute of effect.

The above provisions, although seemingly adequate for an incipient world space programme, do not adequately address modern exigencies of outer space activity such as collaboration in space stations where repair missions and salvage activities may call for multinational crews, joint space exploration calling for multiple space technology, and transportation to outer space of passengers.

---

[8] Cheng (1997), p. 458.

[9] Like the earlier cited Treaty provision, this provision is derived from United Nations documentation and has been reproduced almost verbatim from paragraph 7 of the 1963 General Assembly Declaration appearing in Resolution 1962 (XVIII). The Treaty provision extends the scope of application of the provision to conduct of astronauts both inside and outside the spacecraft.

[10] Cheng (1997), p. 459.

[11] Agreement Governing the Activities of States on the Moon and other Celestial Bodies, UN Doc A/RES/34/68 of 5 December 1979.

Conduct of the Space Tourist

The recognition that the scope of manned space flight is being expanded from the flight of astronauts to other persons such as repair crew and passengers, is becoming evident. In an attempt in 1988 at drafting a Convention on manned space flight, a team of draftsmen comprising a distinguished cluster of experts in space law from Germany, the Russian Federation and the United States succeeded in a sustained attempt at producing a draft legal instrument which covers certain exigencies of personal conduct in space travel. The most significant thrust of this draft Convention is that it blends harmoniously the essential qualities of scholarship and practicality. The draft Convention has been published in order to draw the attention of the world community of space lawyers and seek comments. It effectively conveys the fundamental postulate that manned space flight is the cornerstone of exploration of outer space and therefore its development requires guidelines on international cooperation and liability.

The draft Convention, in Article III links itself to the Outer Space Treaty principle of awarding jurisdiction in relation to a manned space object and person therein to the State of registry in relation to occurrences in outer space or in a celestial body or on or in the high seas or any other place beyond the jurisdiction of any State. Article IV of the draft Convention devolves responsibility and authority over a manned space flight, the space object involved in such flight and all persons on board, on the commander of the space object. The commander is given sole authority throughout the flight to use any reasonable and necessary means to achieve this end. The same provision makes both the commander and all members of the crew answerable to a person identified as the Director of Manned Space Flight Operations, who is defined by the draft Convention as a person who is designated by the State exercising jurisdiction and control over the space object to be in charge of a particular manned space flight. By this measure, the draft Convention skilfully and unequivocally identifies the chain of command, giving the commander absolute authority on the spot over all those in the space object during the flight, while making him answerable to a person designated as Director of Manned Space Flight Operations, who presumably will be on ground and in mission control.

The draft Convention also ensures safety of persons involved in a manned space flight whether they be crew, passengers or any other category of persons affected by such flight. One of the strengths of the draft Convention is its provision with regard to environmental pollution or other harm caused to the environment by a manned space flight, where the instrument lays responsibility on States whose manned space exploration may jeopardise an existing environmental balance. It also provides for assistance to be given by persons in a manned space flight to others in distress in outer space and prescribes international responsibility on States, whether the space flights in question are carried out by governmental or non governmental entities.

Outer space and celestial bodies can be used for the common heritage of mankind but are *res extra commercium* like the high seas. However, here the distinction ends, in that unlike the high seas, which can be appropriated in certain circumstances, such as through acquiescence by one State of appropriation of an area of the high seas by another, outer space or celestial bodies cannot be appropriated under any circumstances. It is not possible to apply the principle of appropriation to the conduct of crews of other persons in outer space. One cannot establish a pattern of conduct as a prescriptive right in outer space because there are no territorial limits demarcated by and between individual States in outer space. *A fortiori*, outer space has been identified as one composite area which cannot be appropriated by one particular State to the exclusion of others by Treaty provision.

Freedom of outer space, which lays the foundation for conduct of persons in outer space, is enshrined in Article 1 of the Outer Space Treaty of1967, which stipulates that the exploration and use of outer space, including the moon and other celestial bodies, shall be carried out for the benefit and in the interests of all countries, irrespective of their degree of economic or scientific development, and shall be the province of all mankind. The provision also requires outer space to be free for exploration and use by all States without discrimination of any kind, on a basis of equality and in accordance with international law. Finally, the provision grants free access to all States in relation to all areas of celestial bodies.

On a purely superficial comparison of the freedoms of outer space and the high seas, one can notice a general similarity in that both areas are open to mankind equally. However, the purposes for which the areas can be used are intrinsically different. For example, space law is all encompassing on the subject of the conduct of humans in outer space and celestial bodies. The Outer Space Treaty makes the sweeping statement that outer space and celestial bodies shall be open for exploration by mankind, which includes, *inter alia*, the freedom to conduct research, experiments and other forms of exploration. The Convention on High Seas, on the other hand is inclusive and therefore restrictive in forming areas of specific activity. The Convention on the Law of the Sea of 1982 has somewhat remedied this *lacuna* by adding, in Article 87(1), the freedom of scientific research *inter alia* to the already existing four freedoms.

Criminal conduct is an area where the principle of international law applicable to the High Seas lend themselves as a useful analogy to space law. Of course, the offence of piracy cannot be committed by astronauts who are sent to outer space in spacecraft belonging to a State. The offence has to be committed for private ends by persons in a private ship or craft. The offence of piracy in the high seas would nonetheless apply as an analogy to a similar offence committed by private individuals in outer space who do not represent a State as official crew members. This would cover the improbably but nonetheless possible events of the future such as a mutiny on board a commercial spacecraft carrying passengers (which is an analogy derived from shipping law). Piracy in outer space may also occur in instances where personnel of a space craft could act on the orders of a recognized government which is in gross breach of international law and which show a criminal disregard for human life.

Of direct analogy to the draft Convention are three Conventions and a Declaration at air law which lay down principles of law with regard to criminal conduct of persons on board aircraft. Of course, understandably, given the fact that crimes on board aircraft have been prolific over the past years in comparison, these instruments are somewhat more evolved than the space law initiative and have concentrated on such offences as hijacking and other terrorist acts. But they also provide for sanctions against general criminal conduct and therefore would form a useful paradigm for future space law initiatives, particularly on the subject of terrorism in outer space and celestial bodies. The Tokyo Convention of 1963 on Offenses and Certain Other Acts Committed on Board Aircraft,[12] refers to any offence committed or act done by a person on board any aircraft registered in a contracting State, while the aircraft is in flight or on the surface of the high seas or of any other area outside the territory of such State.[13] The aircraft is considered to be in flight from the moment power is applied for the purpose of take off until the moment when the landing run ends.[14] In addition, the Tokyo Convention mentions acts of interference, seizure of or other wrongful exercise of control of an aircraft, implying its concern over hijacking.[15]

The Hague Convention of 1970[16] in Article 1 identifies any person who, on board an aircraft in flight, unlawfully by force or threat or by any other form of intimidation seizes or takes control of such aircraft, or even attempts to perform such an act, as an offender.[17] Anyone who aids such an act is an accomplice, and is included in the category of the former.[18]

It is clear that the Hague Convention by this provision has neither deviated from Article 11 of the Tokyo Convention nor offered a clear definition of the offence of hijacking. It merely sets out the ingredients of the offence – the unlawful use of force, threat or any other form of intimidation and taking control of the aircraft. The use of physical force, weapons or firearms or the threat to use such modes of force are imputed to the offence in this provision. The words force, threat or intimidation indicate that the element of fear would be instilled in the victim. It is an interesting question whether these words would cover an instance where the use of fear as an implement to execute the offence of hijacking covers non-coercive measures such as the drugging of food or beverages taken by the passengers or crew. The Hague Convention does not ostensibly cover such instances. In this context, many recommendations have been made to extend the scope of its Article 1.[19] It is also interesting that the Convention does not envisage an instance where the offender

---

[12]Hereafter referred to as the Tokyo Convention.

[13]Tokyo Convention, Chapter 1, Article 1(2).

[14]Tokyo Convention, Chapter 1, Article 1(3).

[15]Tokyo Convention, Chapter 4, Article 11.

[16]*Convention for the Suppression of Unlawful Seizure of Aircraft*, the Hague, 16 December 1970, hereafter referred to as the Hague Convention.

[17]*Convention for the Suppression of Unlawful Seizure of Aircraft*, the Hague, 16 December 1970, Hague Convention, Article 1(a).

[18]*Convention for the Suppression of Unlawful Seizure of Aircraft*, the Hague, 16 December 1970, Hague Convention, Article 1(b).

[19]See Shubber (1973), p. 687 at 692.

is not on board the aircraft but remains on ground and directs operations therefrom after planting a dangerous object in the aircraft. According to Article 1, the offence has to be devoid of a lawful basis albeit that the legality or illegality of an act is not clearly defined in the Convention.

It is also a precondition in Article 1 that the offence has to be committed in flight, that is while all external doors of the aircraft are closed after the embarkation of the passengers and crew.[20] The mobility of the aircraft is immaterial. Furthermore, Article 1 is rendered destitute of effect if an offence is committed while the doors of the aircraft are left open.

The Tokyo Convention was perhaps the first major attempt at curtailing the menace of hijacking. Not only did it deal solely with jurisdiction over offenses committed in an aircraft in flight, but it also did not exclude any criminal jurisdiction which would have been exercised according to the provisions of any law.[21] The element of nationality underlines the parochial nature of the treatment of the offence and the obstinate refusal of the international community to infuse a universality to the treatment of the offence. Perhaps, as one commentator observed[22] the international community was not prepared in 1963 to address this problem on a collective basis.

The subsequent Hague Convention emphasises that each contracting State undertakes to impose severe penalties without defining what these penalties should be.[23] Furthermore, the geographic limitations set out in Article 1 curtails the punitive measures recommended in the Convention significantly. The Convention makes a further serious omission in stating that it applies only if the place of take off and place of landing is outside the State of registration of the aircraft.[24] This gives rise to a serious anomaly in that if an aircraft with a destination outside its territory of nationality is seized in mid air prior to leaving its airspace and brought back to the place of take off the Convention would not apply even to scheduled flight.

The Montreal Convention which followed in 1971, although extending the period in which the offence would be committed to a period beginning at pre flight preparation which ends twenty four hours after landing, does not cover acts of sabotage, destruction or any damage effected before the above period starts after 24 hours after landing. The Montreal Convention however, is the best attempt so far at attempting to control or curb the offence of hijacking on an international level. Even so, the three attempts so far at international accord fail to cover certain gaps which still exist in this area of prevention and control of hijacking. They are:

---

[20]*Convention for the Suppression of Unlawful Seizure of Aircraft*, the Hague, 16 December 1970, Hague Convention, Article 3, para 1.

[21]Tokyo Convention, Chapter 4, Article 3(3).

[22]Lowenfeld (1972), p. 87.

[23]*Convention for the Suppression of Unlawful Seizure of Aircraft*, the Hague, 16 December 1970, Hague Convention, Article 2.

[24]*Convention for the Suppression of Unlawful Seizure of Aircraft*, the Hague, 16 December 1970, Hague Convention, Articles 1 (4), 3, 4, and 7.

(a) That the Conventions do not provide for and guarantee the trial of an offender and do not specify adequate punitive measures;
(b) That no obligation is cast on contracting States for the extradition of an offender;
(c) That no provision is made for the universal adoption of standards of precaution and safety; and,
(d) That the initial attempt, *albeit* somewhat unsophisticated, of the Tokyo Convention at a remedial approach has been thwarted by the repressive attitude of the two subsequent Conventions.

The Bonn Declaration of 1978 was yet another attempt by the international community to combat terrorism related to international civil aviation. The major economic powers Canada, France, Federal Republic of Germany (as it then was), Italy, Japan, UK and USA collaborated to intensify efforts to combat terrorism. The seven signatory States pledged to take immediate action to cease all flights to a country which refuses to extradite hijackers or return hijacked aircraft, and to halt all incoming aircraft from that country or from any airlines of that country.

The Convention for the Suppression of Unlawful Acts Against The Safety of Civil Aviation signed at Montreal on 23 September 1971,[25] also fails to define in specific terms, the offence of hijacking, although it circumvents barriers placed by Article 1 of the Hague Convention.[26] For instance : it encircles instances where an offender need not be physically present in an aircraft; includes instances where an aircraft is immobile; its doors open; and even draws into its net any person who disseminates false information which could endanger an aircraft in flight.[27] None of the three conventions have however succeeded in identifying the offence of hijacking or advocating preventive measures against the offence itself.

References

Cheng C (1997) Studies in International Space Law. Clarendon Press, Oxford, p 259 and 460

Frangialli F (1999) Crystal ball gazing, dossier. The Courier, 2

Lowenfeld AF (1972) Aviation law, cases and materials. M. Bender, New York, 87

Schenkman J (1955) International civil aviation organization. Librairie E. Dross, Geneve, pp 120–121

Shubber S (1973) Aircraft hijacking under the Hague convention 1970 – a new regime? Int Comp Law Q 22:692

[25]Hereafter referred to as the Montreal Convention.

[26]Montreal Convention, Article 1.

[27]Montreal Convention, Article 1.

Chapter 5
Legal and Regulatory Regime

The view has however been forwarded that under the principles of the Bonn Declaration, retaliatory action in the nature of self preservation by third States against an offending State are subjectively assessed, and that action taken should be commensurate with the gravity of the infringement of the provisions of the Declaration.[1] Be that as it may, the Bonn Declaration was a clear demonstration that positive action can be taken against the threat of interference with international civil aviation if there is international cooperation. The signatories to the Bonn Declaration were the largest manufacturers of aircraft and their airlines covered more than half the world's scheduled passenger kilometres. They obviated the insurmountable difficulty of obtaining all the signatures of the ICAO contracting States while amply demonstrating that agreements of this nature are effective tools against the terrorist.

While there is a relatively well articulated legal regime governing activities of States and private entities,[2] including a regulatory arm in the form of the Committee on the Peaceful Uses of Outer Space (COPUOS) which is a United Nations body with no law making or adjudicatory powers,[3] it has been said that the involvement of ICAO in space tourism should also be seriously considered.[4] *Jakhu* and *Battacharya* anchor their argument in support of ICAO involvement in regulating outer space traffic on the fact that space traffic will have to be harmoniously blended with air traffic as space and air traffic management issues will be a major factor in the allocation of "air slots" to both aircraft and spacecraft. They envision ICAO's role to be critical in the near future as a regulator and arbiter resolving space traffic issues.[5] *Jasentuliyana*'s thinking on the regulation of outer space activities, published in 1995, introduces another dimension where he suggests that COPUOS could draw on the work of ICAO:

> ...COPUOS could, by following the example of some of the specialized agencies of the United Nations, like the World Health Organization (WHO), the International Maritime

---

[1]Schwenk (1979), p. 307 at 321.

[2]See Lachs (1980), p. 12.

[3]See Steinhardt (1995), p. 753 at 757.

[4]Jakhu and Battacharya (2002), pp. 112 et seq.

[5]Jakhu and Battacharya (2002), pp. 112 et seq.

R. Abeyratne, *Space Security Law*,
DOI 10.1007/978-3-642-16702-7_5,

Organization (IMO) and particularly the International Civil Aviation Organization (ICAO) seek to formulate international standards and recommended practices.[6]

Wassenbergh attempted in 1997 to answer the question as to whether there would be a need for a new global space organization to monitor the public law aspects of space activities or whether ICAO could act as such.[7] Although he did not answer the question directly, he offered the conclusion that regulation by a space organization would ensure the safety of the activities, orderly international competition and optimum protection of the Earth and space environment.[8]

Whether the existing regime within COPUOS regulates space tourism or whether a separate Organization is created, or indeed whether regulation of space tourism is brought within ICAO will be decided as a matter of practicality. Be that as it may, it is incontrovertible that some ICAO involvement will be necessary, in view of the inevitable overlap between the air transport and space transport segments of a journey, particularly in the field of air traffic management. Therefore, the practical approach to the fusion of regulation between the air transport and outer space travel could lie in extending the regulatory regime falling under the umbrella of the Chicago Convention.[9] However, as *Jasentuliyana* suggests, the existing outer space activity regime, in terms of regulation could remain within COPUOS with analogous principles attenuated from the Annexes to the Chicago Convention on subjects such as licencing, documentation to be carried on board, and rules of outer space travel. These principles should address the need for some accountability and responsibility on the part of States with regard to Standards and Recommended Practices, as enunciated in Articles 37 and 38 of the Chicago Convention.[10]

A good starting point with regard to the relevance of ICAO to space tourism in particular and space travel in general is to look at the current applicability of the Chicago Convention. The primary consideration is the status of a sovereign state to launch space tourism. The principle of State sovereignty in airspace is embodied in Article 1 of the Chicago Convention which recognizes that every State has

[6]Jasentuliyana (1995), p. 349 at 380.

[7]Wassenbergh (1997), pp. 529–535.

[8]Wassenbergh (1997), p. 535.

[9]Convention on International Civil Aviation, signed on 7 December 1944 at Chicago. See ICAO Doc 7300 (Ninth Edition) 2006.

[10]Article 37 requires each contracting State to undertake to collaborate in securing the highest practicable degree of uniformity in order to improve and facilitate air navigation. To this end, ICAO is charged in the same article, with adopting and amending from time to time, international standards and recommended practices and procedures dealing with 11 key areas, including communications systems, licencing of operating and mechanical personnel, airworthiness of aircraft, registration and identification of aircraft log books, aeronautical maps and charts... and any such other matters concerned with the safety, regularity and efficiency of air navigation, as may from time to time, seem appropriate. Article 38 of the Convention obliges contracting States to notify ICAO of any differences between their own regulations and practices and those established by international standards or procedures. The notification of differences however, does not absolve States from their continuing obligation under Article 37 to collaborate in securing the highest practicable degree of uniformity in international regulations, standards, and procedures.

sovereignty over the air space above its territory, the latter being defined in Article 2 as land situated within and water adjacent to the State concerned. As for rights over airspace over the high seas, Article 87 of the United Nations Convention on the Law of the Sea of 1982[11] awards freedom for the aircraft of all States to fly over the high seas. An important consideration in delineating territorial sovereignty lies in the expansion of Flight Information Regions (FIR) and the provision of air traffic management services by States particularly when such measures are influenced by the revenue generating capabilities that are inherent in such an expansion of scope. The Chicago Convention, in its vision and wisdom, incorporates various provisions regarding the provision of air navigation services by States to aircraft flying over their territories. Firstly, the Convention guarantees, through provisions included in Chapter XV, that States, which are unable to provide air navigation services to aircraft will be assisted. Secondly, Article 15 of the Convention assures airlines that every airport in a Contracting State that is open to public use by its national aircraft shall also be open under uniform conditions to the aircraft of all the other Contracting States. The conditions are deemed to apply to the use, by aircraft, of every Contracting State of all air navigation facilities, including radio and meteorological services, which may be provided for public use for the safety and expedition of air navigation. Charges levied for such services are deemed by Article 15 to be anti-discriminatory whereby aircraft are not to be charged for airports and air navigation services provided to them at a rate higher than those levied on the national carrier of the State which provides the service. To this end, Article 28 of the Convention obligates Contracting States to provide, as far as practicable in their territories, airports, radio services, meteorological services and other air navigation facilities to facilitate international air navigation according to Standards established pursuant to the Convention.

The tightly-set legal parameters of the Chicago Convention, particularly the assurance of air navigation services on an equal and non-discriminatory basis, are relevant in the 21st Century, where service providers and airline operators have to collaborate in ensuring a seamless global air navigation system. Modern technology offers sophisticated air-ground data communications by VHF (very high frequency) and satellite, assisted by precise navigation by inertial/GNSS and computing in air traffic services. These will be used in the negotiation of "dynamic user preferred routes" offering various alternatives to airline operators which provide fuel and time savings. However, such preferences for flight profiles and uses thereof will be subject to meteorological exigencies which have to be cautiously assessed. This imposes an added burden on both the service provider and airline operator. Judgment and interpretation will be critical factors in this process, an inevitable corollary of which will be the need to examine legal aspects of the modern seamless air traffic management system.

---

[11]The Law of the Sea, Original Text of the *United Nations Convention on the Law of the Sea*, all Annexes and Index, United Nations: New York, 1983.

As stated earlier, responsibility of States for the provision of air navigation services in their territories is founded in principles contained in Article 28 of the Chicago Convention of 1944.[12] It must be noted that this is not an absolute obligation as the State is called upon to provide such services only in so far as it finds practicable to do so. In order to cover an eventuality of a State not being able to provide adequate air navigation services, the Convention imposes an overall obligation on the Council of ICAO in Article 69 to consult with a State which is not in a position to provide reasonably adequate air navigation services for the safe, regular, efficient and economical operations of aircraft. Such consultations will be with a view to finding means by which the situation may be remedied. Article 70 of the Chicago Convention even allows for a State to conclude an arrangement with the Council regarding the financing of air navigation facilities and the Council is given the option in Article 71 of agreeing to provide, man, maintain and administer such services at the request of a State.

The 31st Session of the ICAO Assembly, in December 1995, adopted Resolution A31-7 pursuant to which the Council established a Panel of Experts on the Establishment of a Legal Framework with Regard to Global Navigation Satellite Systems (LTEP). The Panel was charged, *inter alia*, to develop a legal framework of conduct regarding the use of the GNSS. The framework developed by the LTEP took the form of a Charter containing various principles for the implementation and operation of GNSS, such as recognition of the paramount nature of safety in international civil aviation; non-discrimination and universal applicability and accessibility of GNSS; inviolability of States' sovereign rights; continuity, integrity, availability and reliability of services; and international cooperation. The Charter was recognized by the ICAO Council, at its 153rd Session in March 1998, as being worthy of submission to the 32nd Session of the Assembly, which in turn adopted the text of the Charter in a Resolution. The ICAO Assembly, at its 32nd Session in 1998, adopted Resolution A32-19 (Charter on the Rights and Obligations of States Relating to GNSS Services) containing fundamental principles of a Charter of Rights and Obligations of States in relation to GNSS Services.[13] In the Resolution, States, whilst recognizing that the primary use of GNSS services is to maintain safety in international civil aviation, reaffirm the principle that every State and aircraft of all States shall have access, on a non-discriminatory basis, and under uniform conditions to the use of GNSS services. The Resolution also grants every State authority and control over aircraft operations over their territory, and, *inter alia* imposes obligations on provider States to ensure continuity, availability, integrity, accuracy and reliability of such services.

---

[12]The Law of the Sea, Original Text of the *United Nations Convention on the Law of the Sea*, all Annexes and Index, United Nations: New York, 1983, Article 28 a.

[13]Res A32-19, Charter on the Rights and Obligations of States Relating to GNSS Services. *Assembly Resolutions in Force* (as of 28 September 2007) Doc. 9902, at V-7.

It is at this point that the problem arises, and legal discourse begins. The mere fact that the Charter is now an ICAO Assembly Resolution has prompted the comment:

> Adopted in the form of an Assembly Resolution, the Charter cannot be accorded any legal force and therefore must be regarded as legally not binding. Some commentators, having expressed serious doubts as to the usefulness of the instrument, seem to be somewhat displeased with the nomenclature employed which would be indicative of a legal instrument of fundamental importance[14]

Another view on the ICAO Assembly states that:

> [ICAO] Assembly recommendations ... are more than hortatory. They are designed to set global norms in a field where there is widespread acknowledgment of the need for ordered conduct. They are adopted by a plenary body, with the shared expectation that States will follow them to the extent that they are able to. They clearly are not binding, but they have a sufficient channelling effect to place them well above the low point on a continuum of normative instruments ranging from non law to true law.[15]

Unlike the former view, this statement attributes more coercive force to ICAO Assembly Resolutions, thus making the Charter on States' rights and obligations relating to GNSS an instrument which sets out norms and demands States to follow them if possible and is therefore in between "non law" and "true law." What this means is that the Charter would not be totally destitute of effect in establishing certain obligations for States to perform. Therefore, the Charter becomes a reckonable force in international relations if not at international law, particularly since ICAO resolutions are highly persuasive and carry much political leverage. Above all, such an instrument could, while reaffirming existing legal principles, pave the way for an international convention that is binding on States' Parties.

The 32nd Session of the ICAO Assembly also adopted Resolution A32-20 (Development and Elaboration of an Appropriate Long Term Legal Framework to Govern the Implementation of GNSS). This resolution, which recognizes that GNSS is an important element of the ICAO CNS/ATM System, is aimed at providing a framework governing safety-critical services for aircraft navigation with worldwide coverage. It also recognized, *inter alia*, the need for an appropriate long term legal framework to govern the implementation of GNSS and endorsed the Council decision taken earlier to authorize the ICAO Secretary General to establish a Study Group on Legal Aspects of CNS/ATM systems. The Assembly instructed the Council and the Secretary General to consider the elaboration of an appropriate long term legal framework to govern the operation of GNSS including consideration of an international convention.

Resolution A32-20 was a signal development in global recognition of the need to inquire into a liability regime regarding damage caused by the GPS process through signals transmitted to air navigation facilities. The Resolution resulted in an ICAO Secretariat Study Group being established to elaborate proposals for a liability

---

[14]Andrade (2001), p. 89.

[15]Kirgis Jr. (1995), p. 825 at 840.

framework. The Study Group reported to the 33rd Session of the Assembly (held in September-October 2001) in a somewhat divided way, some members recommending that the applicable regime under domestic law was adequate and appropriate to cope with the global navigation satellite system and others being of the view that a global international law instrument, such as a Convention, might be required to address issues of liability in the long term. A compromise between these two views, representing a model contractual framework, was also suggested to the Assembly as an alternative. The Assembly further remanded the matter to the Study Group to finalize the concept of a contractual framework as a first step, with a view to later considering the development of an international convention as a long term measure.

The transformation of a space flight from the air transport segment to the outer space segment brings to bear the need to sustain the communications process between the service providers and the users as a fluid extension of the process which was initially carried out during the air transport stage. Since the Chicago Convention applies exclusively to civil aviation and civil aircraft, there is little point in amending the Convention to cover activities in outer space. The pragmatic way would be to construct a space law regime that would address the issues involving space tourism while attenuating analogous principles from the Chicago Convention and other legal instruments of air law which have already been cited. Cross references where necessary to the relevant air law instruments could be made in such a regime where there are overlaps in issues between the air and space segments in a journey, particularly involving the conduct of passengers, liability for damage caused by space debris, and communications concerned with flight.

As an initial measure, experts of both ICAO and COPUOS could form a think tank with a view to identifying the issues that need to be addressed. ICAO has a history of collaborating well with international organizations, such as the International Maritime Organization and the International Telecommunications Union on transportation issues. However, it would be a challenge for ICAO to view the existing provisions of the Chicago Convention in the light of the extended perspectives of space flight. The Convention could be amended where necessary, in accordance with Article 94 which prescribes that a two-thirds majority of the ICAO Assembly could make treaty amendments. The pre-existence of a comprehensive aviation regulatory structure makes it somewhat easier for the aviation experts to harmonize existing structure to ease into a fused air-space structure.

From the space angle, the space lawyers would be primarily concerned with the existing principles and their adequacy. Two conventions, the Outer Space Treaty of 1967[16] and the Liability Convention of 1972.[17] Would be starting points. Both Conventions impose prima facie an obligation on States under *jus cogens* or generally endorceable and applicable law. If it can be accepted that a principle of

---

[16]Treaty on Principles Governing the Activities of States in the Exploration and Use of Outer Space, including the Moon and other Celestial Bodies Opened for Signature at Moscow, London and Washington on 27 January 1967. 610 UNTS 205.

[17]Convention on the International Liability for Damage Caused by Space Objects, Opened for Signature at London, Moscow and Washington on 29 March 1972. 961 UNTS 187.

jus cogens creates obligations *erga omnes*, it becomes an undeniable fact that Article 1(1) of the *Outer Space Treaty* could be considered a peremptory norm or *jus cogens,* since it generates obligations towards the international community as a whole. *Christol* observes:

> Article 1 Paragraph 1 of the Space Treaty, with its adoption of the common benefits and interests guarantee, can be supported (as an example of peremptory norms) because the provisions conform to moral law in the sense that all humankind is to benefit unconditionally, and because the terms are consistent with the spirit and the purposes identified in Article 1 Pars. 1 through 3 and Article 2 Pars. 1 through 4 of the UN Charter, as well as with complimentary international agreements of lesser authority. To the extent that the terms are beneficial to individuals, the larger community, and States, and when the provisions are found on the fundamental moral principles contained in the foregoing paragraphs of Article 1 and 2 of the UN Charter, such basic principles qualify for the status of peremptory norms of general international law.[18]

The effect of this observation is that the content and nature of Article 1 (1) confirms that it is a *jus cogens*. There is seemingly no reason why the international community should not give such recognition to the "common interest" principle as enshrined in Article 1(1) which is aimed at the protection of the interests of the international community as a whole. *A fortiori*, on the same basis, Article IX of the *Outer Space Treaty* which requires that States should avoid harmful contamination and adverse change in the environment of the Earth which may result from the exploration of outer space would incontrovertibly be considered *jus cogens*.

Article VI of the *Outer Space Treaty* provides in part that State Parties to the Treaty shall bear international responsibility for national activities in outer space, whether such activities are carried out by governmental agencies or non-governmental agencies. This provision clearly introduces the notion of strict liability *erga omnes* to the application of the *jus cogens* principle relating to outer space activities of States and could be considered applicable in instances where States hold out to the international community as providers of technology achieved and used by them in outer space, which is used for purposes of air navigation. Article VI further requires that the activities of non-governmental entities in outer space shall require authorization and continuing supervision by the appropriate State Party to the Treaty, thus ensuring that the State whose nationality the entity bears would be vicariously answerable for the activities of that organization, thereby imputing liability to the State concerned.

Article VII makes a State Party internationally liable to another State Party for damage caused by a space object launched by that State. The *Registration Convention* of 1974[19] in Article II(1) requires a launching State of a space object that is launched into earth orbit or beyond, to register such space object by means of an entry in an appropriate registry which it shall maintains and inform the Secretary General of the United Nations of the establishment of such a registry. This

[18]Christol (1983), p. 6.

[19]*Convention on Registration of Objects Launched into Outer Space*, adopted by the General Assembly of the United Nations, New York, 12 November 1974, 1023 UNTS 15.

provision ensures that the international community is kept aware of which State is responsible for which space object and enables the United Nations to observe outer space activities of States. Article VI of the Convention makes it an obligation of all State Parties, including those that possess space monitoring and tracking facilities, to render assistance in identifying a space object which causes damage to other space objects or persons. Justice Manfred Lachs analyses these provisions of the *Registration Convention* to mean that the State of registry and the location of the space object would govern jurisdictional issues arising out of the legal status of space objects.[20] On the issue of joint launching of space objects, Justice Lachs observes:

> No difficulties arise whenever a State launches its own object from its own territory; the same applies to objects owned or launched by non-governmental agencies registered in that State. However, in cases of joint launching, agreement between the parties is required as to which of them is to be deemed the "State of Registry." A similar agreement is also necessary when a launching is carried out by an international organization.[21]

The above provision ensures the identification of parties responsible for specific activities in outer space and thereby makes it easier to impose liability for environmental damage caused.

The *Outer Space Treaty*,[22] while expostulating the fundamental principle in its Article 1 that the exploration and use of outer space, including the moon and other celestial bodies, shall be carried out for the benefit and in the interests of all countries, explicitly imposes in Article VII international liability and responsibility on each State Party to the Treaty, for damage caused to another State Party or to its populace (whether national or juridical) by the launch or procurement of launch of an object into outer space. In its preceding provisions the Treaty imposes international responsibility on States Parties for national activities conducted in outer space. The Treaty also requires its States Parties to be guided by the principle of co-operation and mutual assistance in the conduct of all their activities in outer space.[23] This overall principle is further elucidated in the same provision:

> States Parties to the Treaty shall pursue studies of outer space, including the moon and other celestial bodies, and conduct exploration of them so as to avoid harmful contamination and also adverse changes in the environment of the Earth resulting from the introduction of extra terrestrial matter.[24]

[20] Lachs (1972), p. 70.

[21] Lachs (1972), p. 70.

[22] *Treaty on Principles Governing the Activities of States in the Exploration and Use of Outer Space, including the Moon and Other Celestial Bodies*, opened for signature at Moscow, London and Washington, 27 January 1967, 610 UNTS 205.

[23] *Treaty on Principles Governing the Activities of States in the Exploration and Use of Outer Space, including the Moon and Other Celestial Bodies*, opened for signature at Moscow, London and Washington, 27 January 1967, 610 UNTS 205. Article IX.

[24] *Treaty on Principles Governing the Activities of States in the Exploration and Use of Outer Space, including the Moon and Other Celestial Bodies*, opened for signature at Moscow, London and Washington, 27 January 1967, 610 UNTS 205. Article IX.

The *Moon Agreement*[25] of 1979 provides that in the exploration and use of the moon, States Parties shall take measures *inter alia* to avoid harmfully affecting the environment of the earth through the introduction of extra terrestrial matter or otherwise.[26]

The *Liability Convention*[27] contains a provision which lays down the legal remedy in instances of damage caused by Space objects. Article II provides:

> A launching State shall be absolutely liable to pay compensation for damage caused by its space objects on the surface of the Earth or to aircraft in flight,[28]

thereby imposing a regime of absolute liability on the State that launches space objects, from satellites which provide technology and communication that is used for air navigational purposes, to spacecraft carrying passengers. Although admittedly, both the *Outer Space Treaty* and the *Liability Convention* do not explicitly provide for damage caused by technology and communication provided by space objects, culpability arising from the "common interest" principle and liability provisions of the two conventions can be imputed to States under these Conventions.

Gorove states that in the field of international space law, two clearly connected terms have been used: liability and responsibility.[29] Although "responsibility" has not been cohesively interpreted in any legal treaty relating to outer space, "liability" occurs in the *Liability Convention* and is sufficiently clear therein. This, however, does not mean that State responsibility is not relevant to the obligations of States as, in international relations, the invasion of a right or other legal interest of one subject of the law by another inevitably creates legal responsibility. Therefore, the challenge of determining a suitable regime for space tourism would indeed be a daunting one.

Comparison Between Space Law and Maritime Law

Outer space and celestial bodies can be used for the common heritage of mankind but are *res extra commercium* like the high seas. However, here the distinction ends, in that unlike the high seas, which can be appropriated in certain circumstances,

[25]Agreement Governing the Activities of States on the Moon and other Celestial Bodies, signed on 5 December 1979, UN Doc A/RES/34/68 of 5/12/1979.

[26]Agreement Governing the Activities of States on the Moon and other Celestial Bodies, signed on 5 December 1979, UN Doc A/RES/34/68 of 5/12/1979. Article 7.

[27]*Convention on International Liability for Damage Caused by Space Objects*, March 29 1972, 24 U.S.T 2389, T.I.A.S No. 7762.

[28]Article II(a) defines damage as including loss of life, personal,injury or other impairment of health; or loss or damage to property of States or of persons natural or juridical, or property of international governmental organizations.

[29]Gorove (1983), p. 373.

such as through acquiescence by one State of appropriation of an area of the high seas by another, outer space or celestial bodies cannot be appropriated under any circumstances. In the *Fisheries* case the International Court of Justice adopted the view that a consistent course of action on the part of fishermen of Norwegian vessels had established a pattern at fishing in a certain area over a period time, which had been tacitly acquiesced by the British (who were contesting the right of Norway to fish in the area on the grounds that it was beyond the designated baseline of Norwegian waters) and that therefore Norway had established its right to continue the practice under the application of the general principles of international law to the specific case. The Court held (referring to Norway's practice of fishing in the disputed area):

> In this connection, the practice of States does not justify the formulation of any general rule of law . . . all that the court can see therein is the application of general international law to a specific case.

It is not possible to apply this principle to the conduct of crews or other persons in outer space. One cannot establish a pattern of conduct as a prescriptive right in outer space because there are no territorial limits demarcated by and between individual States in outer space. *A fortiori*, outer space has been identified as one composite area which cannot be appropriated by one particular State to the exclusion of others by Treaty provision.

Freedom of outer space, which lays the foundation for conduct of persons in outer space, is enshrined in Article 1 of the Outer Space Treaty of1967, which stipulates that the exploration and use of outer space, including the moon and other celestial bodies, shall be carried out for the benefit and in the interests of all countries, irrespective of their degree of economic or scientific development, and shall be the province of all mankind. The provision also requires outer space to be free for exploration and use by all States without discrimination of any kind, on a basis of equality and in accordance with international law. Finally, the provision grants free access to all States in relation to all areas of celestial bodies.

Freedom of the high seas is contained in Article 2 of the Convention on the High Seas of 1958 which provides that since the high seas is open to all nations, no State may validly purport to subject any part of them to its sovereignty. Freedom of the high seas are: the freedom of navigation, the freedom of fishing; the freedom to lay submarine cables and pipelines; and the freedom to fly over the high seas. These freedoms are recognized by the principles of public international law.

On a purely superficial comparison of the freedoms of outer space and the high seas, one can notice a general similarity in that both areas are open to mankind equally. However, the purposes for which the areas can be used are intrinsically different. For example, space law is all encompassing on the subject of the conduct of humans in outer space and celestial bodies. The Outer Space Treaty makes the sweeping statement that outer space and celestial bodies shall be open for exploration by mankind, which includes, *inter alia*, the freedom to conduct research, experiments and other forms of exploration.

The Convention on High Seas, on the other hand is inclusive and therefore restrictive in forming areas of specific activity. The Convention on the Law of the Sea of 1982 has somewhat remedied this *lacuna* by adding, in Article 87(1), the freedom of scientific research *inter alia* to the already existing four freedoms.

Criminal conduct is an area where the principles of international law applicable to the High Seas lend themselves as a useful analogy to space law. Of course, the offence of piracy cannot be committed by astronauts who are sent to outer space in spacecraft belonging to a State. The offence has to be committed for private ends by persons in a private ship or craft. The offence of piracy in the high seas would nonetheless apply as an analogy to a similar offence committed by private individuals in outer space who do not represent a State as official crew members. This would cover the improbably but nonetheless possible events of the future such as a mutiny on board a commercial spacecraft carrying passengers (which is an analogy derived from shipping law). Piracy in outer space may also occur in instances where personnel of a space craft could act on the orders of a recognized government which is in gross breach of international law and which show a criminal disregard for human life.

The offence of piracy at sea and its consequences were succinctly defined by Judge Moore in his dissenting judgment in the 1927 *Lotus* Case.

> In the case of what is known as piracy by law of nations, there has been conceded a universal jurisdiction, under which the person charged with the offence may be tried and punished by any nation into whose jurisdiction he may come. I say "piracy by law of nations," because the municipal laws of many States denominate and punish as "piracy" numerous acts which do not constitute piracy by law of nations, and which therefore are not of universal cognizance, so as to be punishable by all nations. Piracy by law of nations, in its jurisdictional aspects, is *sui generis*. Though statutes may provide for its punishment, it is an offence against the law of nations; and as the scene of the pirate's operations is the high seas, which it is not the right or duty of any nation to police, he is denied the protection of the flag which he may carry, and is treated as an outlaw, as the enemy of all mankind - *hostis humani generis* - whom any nation may in the interest of all capture and punish.

Article 15 of the Convention on the High Seas of 1958 defines the offence of piracy as the following:

a. Any illegal acts of violence, detention or any act of depredation, committed for private ends by the crew or the passengers of a private ship or a private aircraft, and directed:
   - On the high seas, against another ship or aircraft, or against persons or property on board such ship or aircraft;
   - Against a ship, aircraft, persons or property in a place outside the jurisdiction of any State.

2. any act of voluntary participation in the operation of a ship or of an aircraft with knowledge of facts making it a pirate ship or aircraft;
3. any act of inciting or of intentionally facilitating an act described in sub-paragraph (1) or sub-paragraph (2) of this Article.

Mutatis mutandis, this provision would serve well as an analogy and persuasive authority in the event a similar offence committed in outer space or celestial body is examined by a competent court of any jurisdiction.

The Convention restricts the application of the offence of piracy to acts on the high seas or "any place outside the territorial jurisdiction of any state" which means essentially *in contextu* an island or "*terra nullius*" (no man's land). The latter is a fitting analogy for outer space or celestial body which is outside the jurisdiction of any State.

Article 19 of the Convention on the High Seas provides for remedial action and grants the right to any State on the high seas, or in any other place outside the jurisdiction of that State, to seize a pirate ship or aircraft, or a ship taken by piracy and under the control of pirates, and arrest the persons and seize the property on board. The courts of the State which carried out the seizure may decide upon the penalties to be imposed, and may also determine the action to be taken with regard to the ships, aircraft or property, subject to the rights of third parties acting in good faith.

The significance of the provisions of the Convention with regard to the offence of piracy, when related to similar offences which may be committed in outer space, lies in the universal application of the various limits of the offence itself. For example the offence of piracy includes acts of violence, detention or any act of depredation. This effectively brings to bear its relevance to instances such as the take over, detention or illegal deviation of a spacecraft and therefore is quite useful as paradigm legislation for space law.

With regard to principles of tortious liability, both space law and shipping law share a common thread of international responsibility of States. Space law, by virtue of the principle that a State of registry retains jurisdiction in its space object, imputes territoriality to a State in relation to its space craft, space station or other space object. Therefore, any act committed by one State or individual representing a State which would adversely affect such space object would be comparable with an instance where the act of one State affects another in its territorial waters.

Principles of State responsibility for negligent acts leading to reparation are concomitants of substantive public international law. International responsibility relates both to breaches of treaty provisions to which States are bound and also to other infringements of legal duty. Brownlie makes bold to assert that there is no harm in using the term "international tort" to describe the breach of a duty which results in a loss to another State. Judge Huber in the *Spanish Zone of Morocco Claims* case observed:

> Responsibility is the necessary corollary of a right. All rights of an international character involve international responsibility. If the obligation in question is not met, responsibility entails the duty to make reparation.

In the celebrated *Corfu Channel* case, decided in 1947, involved the liability of Albania for mine-layering in her international waters which caused space damage to British mine sweepers. The Court held:

> These grave omissions involve the international responsibility of Albania. The Court therefore reaches the conclusion that Albania is responsible under international law for the explosions which occurred . . . and for the damage and loss of human life which resulted from them.

The above principle will incontrovertibly apply equally to space law for two reasons: both space law and the law related to the Sea are grounded in the principles of public international law; and principles of State responsibility are universal and cannot be applied on different bases with regard to species of international law.

The responsibility of States for the negligent acts of their agents in outer space would be based on the concept of objective responsibility *i.e.*: responsibility of States for those acts committed by their officials or its organs and which acts such officials are bound to perform in the course of their duties. The elements of *faute* or fault on the part of the officials is irrelevant. Tortious liability of the State concerned and responsibility therefor would also entail in instances where the officials or State agencies act beyond their competence. The only necessary element for the imposition of State responsibility is the establishment of the fact that such persons acted with authority derived from the State concerned when the tortious act was performed.

On 29 September 1988, the Governments of the United States, the Member States of the European Space Agency, the Government of Japan and the Government of Canada signed an agreement on co-operation between these States in the detailed design, development, operation and utilization of the permanently manned Civil Space Station. The objective of this Agreement was to establish a long term international collaborative framework among the States' Parties, *inter alia*, on the design and development of a manned space station for peaceful purposes, in accordance with the principles of international law. The Agreement provides for the establishment of the permanently manned Space Station Complex in low Earth orbit which would also comprise unmanned elements.

Article 22 of the Agreement provides that each States' Party shall exercise criminal jurisdiction over the flight elements they respectively provide and over personnel in any flight element who are their respective nationals as registered by each respective States' Party. In addition, the Agreement gives the United States the power to exercise criminal jurisdiction over a non-US national for misconduct, whether such misconduct occurs in a US element or non- US element attached to the Space Complex. Of course, this power has to be exercised by the United States after consultation with the partner State of which the miscreant is a national or after receiving that States' concurrence for the exercise of US jurisdiction. The United States can also exercise its jurisdiction if the partner State does not respond to the United States' request for consultations on an incident involving a non-national of that State, or if such State fails to give its assurance that it would exercise its own jurisdiction on its national, so charged.

A later agreement, signed on 29 January 1998 by and between the Governments of Canada, Governments of Member States of the European Space Agency, Japan, the Russian Federation and the United States concerning co-operation on the Space Station, in Article 22 devolves *in limine* criminal jurisdiction on the States' Parties over personnel in or on any flight element who are their respective nationals. The provision also empowers the States Party whose national affects the life or safety of a national of another States' Party or causes damage to the flight element of another States' Party to discuss prosecutorial issues with the States' Party concerned. The aggrieved States' Party is authorized by Article 22 to exercise criminal jurisdiction over the alleged perpetrator if, within 90 days of consultation, the States' Party of which the perpetrator is a national concurs in such criminal jurisdiction or does not provide the aggrieved States' Party with an assurance that the miscreant will be prosecuted. Article 22 also contains provision with regard to extradition.

Although the nature of outer space and the High Seas and the uses they are put to, are different the underlying principles with regard to the conduct or persons in the two geographic areas are similar, if not identical. Treaty provisions with regard to the use of both are certainly comparable, and to this extent space law could alternate principles from the law of the sea when necessary.

As has been already discussed, proponents of space law should now address the broader issue of persons other than astronauts being involved in space travel in the future. Preparation for this eventuality should not only be made technologically but also intellectually.

References

Andrade AAL (2001) The global navigation satellite system. Ashgate, Aldershot, p 89

Christol CQ (1983) The Jus Cogens principle and international space law. Colloquium 26:6

Gorove S (1983) Liability in space law: an overview. Ann Air Space Law 8:373

Jakhu R, Battacharya R (2002) Legal aspects of space tourism. In: Proceedings of the forty-fourth colloquium on the law of outer space, McGill: Canada, 112 et seq

Jasentuliyana N (1995) A survey of space law as developed by the United Nations. In: Jasentuliyana N (ed) Perspectives on international law, Chap 16. Kluwer Law International, The Hague, p 380

Kirgis FL Jr (1995) Aviation. In: Schachter O, Joyner CC (eds) United Nations legal order, Chap 18, vol 2. Cambridge University Press, Cambridge, 840

Lachs M (1972) The law of outer space, an experience in contemporary law making. Sijthoff, Leiden, 70

Lachs M (1980) The development and general trends of international law in our time. Recueil Des Cours 169:12 (Foreword)

Schwenk W (1979) The bonn declaration in hijacking. Ann Air Space Law 4:321

Steinhardt RG (1995) Outer space. In: Schachter O, Joyner CC (eds) United Nations legal order, vol 2. American Society of International Law, Washington, p 757

Wassenbergh H (1997) An international framework for private space activities. Ann Air Space Law 22(1):529–535

Chapter 6
Issues of Aerospace Insurance

Both sub-orbital flights and the broader issue of space tourism bring to bear the common ground shared by air and space insurance. In recent times, both the air transport and space industries have shown synergies in international legal problems pertaining to the procurement of insurance. This has partially been due to the single most ominous economic throw back from the events of 11 September 2001 concerning the air transport industry and the ensuing insurance crisis. On 17 September 2001, underwriters gave 7 days notice of cancellation of the standard war risk and allied perils clause of the aircraft insurance contract, plunging the commercial airline industry into a causal paradox of necessity and inability in the running of their air services. The resulting gloom, largely stemming from the economic impotence of carriers worldwide in not being able to meet the un-affordable new premium level were somewhat diluted when some States provided financial support to bail out their carriers. However, it became immediately apparent that a certain worldwide and combined effort on the part of nations was necessary if air transport services were to be sustained amidst the crisis.

Now, after nearly 10 years, it is common knowledge that the problems of aviation insurance were gravely aggravated by the events of 11 September 2001, calling for urgent crisis management.[1] A relatively obscure corollary was that the events also indirectly affected space insurance, particularly due to many liabilities incurred and claims received by many insurance underwriters for massive amounts of compensation with respect to the events of 11 September 2001. These claims inevitably reduced the capital set aside by the underwriters for other insurances including space insurance. A corollary to this trend is that some insurers are now focusing more on the lucrative aviation insurance policies that have emerged as a result, due to the sharp rise in premiums for commercial air transportation. Consequently, space insurance has been "shelved" by the underwriters in order for them to concentrate on the correspondingly greater business opportunities offered by commercial air transport insurance.

The commercial trend that has veered the attention of insurance underwriters to commercial air transport insurance has had the further impetus of a series of

[1]See Abeyratne (2002), 595 at pp. 599–634.

R. Abeyratne, *Space Security Law*,

DOI 10.1007/978-3-642-16702-7_6,

communication satellite problems in recent times that has strained the resources of insurers who underwrite space activities. This in turn has imposed a severe strain on companies that launch new space craft. The woes of space insurance are reflected in the figures of the past decade. For instance, in the late 1990s, insurers offered a total coverage to the space industry of $1.3 billion[2] with an exposure of $400 million for a single launch, whereas the total capacity offered by the space insurance industry in 2002, as projected, was as little as $300 million.[3] This figure falls far short of a realistic "cap" on space insurance that could comfort companies hoping to obtain insurance for launches of large geostationary communication satellites. On a basic comparison between the drastic rise in insurance premiums both in the commercial air transport industry and the space industry, the former, after war risk insurance policies were cancelled on 24 September 2001, were raised considerably higher at drastically reduced liability limits,[4] while the latter, in some cases, had rates increased on launch insurance by 50% and for on-orbit insurance for as much as 75%.[5]

The above indicators would seemingly give the perception that space insurance is much worse off than air transport insurance. The reality is that both are in a similar predicament and have the same difficulties in re-surfacing to their *status quo ante*. The problems faced by both insurance industries involve critical risk management which calls for stringent measures to restore the industry to levels that prevailed before 11 September 2001. However, this is not the only factor to be considered. Inasmuch as there are similarities in terms of problems facing both industries, the air transport industry has, unlike the space industry, been given the benefit of a significant boast through the auspices of the International Civil Aviation Organization toward restoring a viable commercial air transport insurance regime. This article will examine space insurance issues as well as air transport insurance issues to identify the common ground experienced by both, with a view to examining a possible approach in order for both industries to survive the current crisis.

The Space Insurance Industry

The space insurance industry became a separate commercial element in the field of insurance in 1965 and the space insurance underwriting community came into being as a result of the rapidly evolving commercial space technologies that called for considerable financial investments. Over the past 3 decades, space insurance underwriters have collected approximately $4.2 billion in premium revenues.

---

[2]Foust (2001), p. 1. See also, Echostar VIII Satellite May Not Be Insured, *Aerospace Risk*, Issue 39, March 2002, at p. 9.

[3]Foust (2001), p. 1. See also, Echostar VIII Satellite May Not Be Insured, *Aerospace Risk*, Issue 39, March 2002, at p. 9.

[4]Abeyratne (2002), at pp. 599–600.

[5]Foust (2001), at p. 3.

Correspondingly, they have paid around $3.4 billion in settlement of claims. The space insurance market has now become a dynamic and highly competitive one, covering from 20 to 30 commercial satellite launches annually.

Similar to most commercial air transport insurance contracts, the space insurance policy is usually underwritten in syndicate where each individual underwriter assumes a percentage of the risk. The coverage of each risk is undertaken for a fractional share of the policy so that the overall risk can be spread out through the global markets. The spreading of risk is accomplished usually through the participation of 10 to 15 large companies and 20 to 30 smaller companies.

One of the most significant and compelling reasons for the predicament faced by the insured in the space industry is the recent rush of communication satellite problems and spacecraft failures. These events have made the space industry a high risk area. The underlying problem, however, is one which afflicts both the space insurance industry as well as the commercial air transport industry in that the contract of insurance in both instances is not regulated on an international basis. The insurance contract in both areas has been exclusively within the realm of the private sector, where the insurance market forces have dictated the fixing of premiums and limits. Insurance of space activities and spacecraft amply reflects the significance of risk management and the space insurance contract primarily plays the role of mollifying investors in a space programme that their investments would be safe and covered by insurance in the event of damage or launch failure.

In the space industry, insurance applies mostly to communication satellites, which have shown a spate of problems in recent times, plunging the "risk factor"[6] of the launch and activity of such spacecraft into critical levels. Furthermore, in the present context, risk management becomes critical for both the insured and the investor in relation to all four types of insurance, i.e. pre-launch insurance; launch failure and initial operation insurance; satellite insurance; and third party liability space insurance.

Pre-launch insurance is a critical area that involves the provision of coverage at the preliminary stage of a space project, from the planning stage, through to the carrying out of the launch.[7] Among possible accidents that may occur at the pre-launch stage that may require insurance coverage are those that may occur in production of the satellite and storage followed by transportation of the satellite from the production site to the launch site.[8] Also critical at this stage is the complex and delicate process of placing the satellite on the launching vehicle.[9]

With regard to launch failure insurance, a critical concern for investors is the possibility of non availability of launch vehicles which would particularly affect

---

[6]Margo generally identifies "risk" as the potential for the occurrence of an uncertain event, and goes on to say that a scientist might define risk as "the continuum or spectrum between uncertainty on the one hand and certainty on the other" . See Margo (1992), p. 79.

[7]For a well compiled account of this subject, see Diederiks-Verschoor (1999), p. 177.

[8]Sgrosso (1993), pp. 187–207.

[9]Sgrosso (1993), pp. 187–207.

investments made by satellite manufacturers. Another risk involved in launch insurance is non placement into orbit as programmed. While the third category, satellite insurance, involves protection against satellite failure in orbit, third party liability space insurance insures against liability arising from damage to a third party during the launch or in-orbit operations of a satellite programme.

Critical to the acquisition of space insurance and the accompanying underwriting process is the value placed on technical information, the role played by the brokers and underwriters, fluctuating market conditions and the various parties concerned. These factors played a crucial role in the Intelsat 708 launch failure, in particular the dissemination of technical information and its role in ensuring insurance claims. It remains to be seen whether judicial interpretation of the value placed on technical information would override the seminal principle established in the 1987 *Martin Marietta* case[10] that a contractual waiver between the parties to an insurance space contract absolving parties from negligence or gross negligence would remain paramount over considerations of tort liability of parties. The significance of the *Martin Marietta* case, which involved Intelsat claiming *Martin Marietta*'s tortuous liability *inter alia* for the failure of one of two satellites launched on Titan III rockets of the latter to reach correct position in orbit, lies in the fact that it establishes the principle in most US jurisdictions that negligence is no longer a sound basis for establishing damages if preceding contractual arrangement or agreement were to preclude such liability. The *Martin Marietta* case followed an earlier case,[11] decided in 1984 and decided partly on contractual liability, where a California Court found that contractual provisions incorporated in an insurance policy or other space insurance contract would absolve a dependant seeking recourse to such contractual waiver against liability and allocation of risk.

Both Conventions impose prima facie an obligation on States under *jus cogens* or generally endorceable and applicable law. If it can be accepted that a principle of *jus cogens* creates obligations *erga omnes*, it becomes an undeniable fact that Article 1 (1) of the *Outer Space Treaty* could be considered a peremptory norm or *jus cogens,* since it generates obligations towards the international community as a whole.

Article VIII of the Outer Space Treaty provides:

> A State party to the Treaty on whose registry an object launched into outer space is carried shall retain jurisdiction and control over such object, and over any personnel thereof, while in outer space or on a celestial body.[12]

---

[10]*Martin Marietta Corporation* v. *International Telecommunication Satellite Organization* (Intelsat), 763 F. Supp. 1327 (D.Md. 1991), 991 F.2d (94th Circ. 1992). See also Mason-Zwaan (1993), pp. 16–24.

[11]*Appalachian Insurance Co.* v. *McDonnell Douglas Corp.,* 214 Cal. App. 3d 1, 262 Cal Rptr. 716 (Cal.App. 4th Dist. 1989).

[12]Like the earlier cited Treaty provision, this provision is derived from United Nations documentation and has been reproduced almost verbatim from paragraph 7 of the 1963 General Assembly Declaration appearing in Resolution 1962 (XVIII). The Treaty provision extends the scope of application of the provision to conduct of astronauts both inside and outside the spacecraft.

However, as Bin Cheng validly points out[13] the interpretation of Article VIII could well result in ambivalence and confusion. The "object" and "personnel" referred to in the Treaty provision do not adequately cover persons who are not "personnel" such as passengers in a spacecraft. Of course, as Cheng maintains, the quasi jurisdiction of the State of registry of the spacecraft can apply both in the instance of conduct in the spacecraft as well as outside the spacecraft on the basis that the astronaut concerned would be deemed to belong to the spacecraft at all times in outer space. Logically, therefore, such jurisdiction could be imputed to passengers, visitors and guests by linking them to the spacecraft in which they travelled. This far reaching generalization would then cover the conduct of an astronaut or other persons while walking on the moon, Mars or other celestial body, as well as such persons who go on space walks outside the spacecraft in which they travelled.

Another provision which sheds some light on past attempts by the international community to identify liability and jurisdictional issues relating to astronauts is Article 12 of the Moon Treaty of 1979[14] which provides:

> States Parties shall retain jurisdiction and control over their personnel, space vehicles, equipment facilities, stations and installations on the moon...

The *Moon Agreement*[15]of 1979 provides that in the exploration and use of the moon, States Parties shall take measures *inter alia* to avoid harmfully affecting the environment of the earth through the introduction of extra terrestrial matter or otherwise.[16]

The *Liability Convention*[17] contains a provision which lays down the legal remedy in instances of damage caused by Space objects. Article II provides:

> A launching State shall be absolutely liable to pay compensation for damage caused by its space objects on the surface of the Earth or to aircraft in flight.,[18]

thereby imposing a regime of absolute liability on the State that launches space objects such as satellites, which provide technology and communication that is used for air navigational purposes. Although admittedly, both the *Outer Space Treaty* and the *Liability Convention* do not explicitly provide for damage caused by

[13]Cheng (1968), 532 at p. 538.

[14]Agreement Governing the Activities of States on the Moon and other Celestial Bodies, UN Doc A/RES/34/68 of 5 December 1979.

[15]Agreement Governing the Activities of States on the Moon and other Celestial Bodies, signed on 5 December 1979, UN Doc A/RES/34/68 of 5/12/1979.

[16]Agreement Governing the Activities of States on the Moon and other Celestial Bodies, signed on 5 December 1979, UN Doc A/RES/34/68 of 5/12/1979, Article 7.

[17]*Convention on International Liability for Damage Caused by Space Objects*, March 29 1972, 24 U.S.T 2389, T.I.A.S No. 7762.

[18]Article II(a) defines damage as including loss of life, personal, injury or other impairment of health; or loss or damage to property of States or of persons natural or juridical, or property of international governmental organizations.

technology and communication provided by space objects, culpability arising from the "common interest" principle and liability provisions of the two conventions can be imputed to States under these Conventions.

Professor Brownlie observes:

> [T]oday, one can regard responsibility as a general principle of international law, a concomitant of substantive rules and of the supposition that acts and omissions may be categorized as illegal by reference to the rules establishing rights and duties. Shortly, the law of responsibility is concerned with the incidence and consequence of illegal acts, and particularly the payment of compensation for loss caused.[19]

International responsibility relates both to breaches of treaty provisions and other breaches of legal duty. In the *Spanish Zone of Morocco Claims* case, Justice Huber observed:

> [R]esponsibility is the necessary corollary of a right. All rights of an international character involve international responsibility. If the obligation in question is not met, responsibility entails the duty to make reparation.[20]

There is also explicit recognition that principles of international law apply to space law. The General Assembly of the United Nations in 1961 adopted the view that international law, including the Charter of the United Nations, applies to outer space and celestial bodies.[21] It is also now recognized as a principle of international law that the breach of a duty involves an obligation to make reparation appropriately and adequately. This reparation is regarded as the indispensable complement of a failure to apply a convention and is applied as an inarticulate premise that need not be stated in the breached convention itself.[22] The ICJ affirmed this principle in 1949 in the *Corfu Channel* Case[23] by holding that Albania was responsible under international law to pay compensation to the United Kingdom for not warning that Albania had laid mines in Albanian waters which caused explosions, damaging ships belonging to the United Kingdom. Since the treaty law provisions of liability and the general principles of international law as discussed complement each other in endorsing the liability of States to compensate for damage caused by space objects, there is no contention as to whether in the use of nuclear power sources in outer space, damage caused by the uses of space objects or use thereof would not go uncompensated. The rationale for the award of compensation is explicitly included in Article XII of the *Liability Convention* which requires that the person aggrieved or injured should be restored (by the award of compensation to him) to the condition in which he would have been if the damage had not occurred. Furthermore, under

---

[19]Brownlie (1990), p. 433.

[20]*1925 RIAA ii* 615 at p. 641.

[21]Resolution 1721 (XVI) adopted on 20 December 1961. See also Article 3 of the *Outer Space Treaty*.

[22]*In Re. Chorzow Factory (Jurisdiction) Case, (1927) PCIJ, Ser. A, no. 9* at p. 21.

[23]*ICJ Reports (1949)*, 4 at p. 23.

the principles of international law, moral damages based on pain, suffering and humiliation, as well as on other considerations, are considered recoverable.[24]

As discussed, both treaty law and general principles of international law on the subject of space law make the two elements of liability and responsibility a means to an end – that of awarding compensation to an aggrieved State or other subject under the law. Therefore, in view of the many legal issues that may arise, the primary purpose of a regulatory body which sets standards on State liability in issues concerning the use of space technology would be to carefully consider the subtleties of responsibility and liability and explore their consequences on States and others involved as they apply to the overall concept of the status of a State as a user of space technology which may cause harm or injury to the latter.

The involvement and responsibility of States in space activities leads to legal accountability of such States for space insurance, and, in this respect, one can discern little difference between the role of States in ensuring that there is provision of insurance coverage for activities in outer space and commercial air transport in an adequate manner.

References

Abeyratne R (2002) Crisis management toward restoring confidence in air transport – legal and commercial issues. J Air Law Commerce 67(3):599–634

Brownlie I (1990) Principles of public international law, 4th edn. Clarendon Press, Oxford, p 433

Cheng B (1968) The 1967 space treaty. J Droit Intl 95:538

Christol CQ (1991) Space law past, present and future. Kluwer Law and Taxation Publishers, Deventer, p 231

Diederiks-Verschoor IH Ph (1999) An introduction to space law, 2nd Revised edn. Kluwer, The Hague, p 177

Foust J (2001) Insurance woes may hurt space industry. Spaceflight Now, 1

Margo RD (1992) Risk management and insurance. Ann Air Space Law 17(1):79

Mason-Zwaan TL (1993) The Martin Marietta case or how to safeguard private commercial space activities. Air Space Law 18(1):16–24

Sgrosso GC (1993) Insurance implications about commercial and industrial activities in outer space. Proceedings of the 36th Colloquium (Graz, 1993), pp. 187–207

[24]Christol (1991), p. 231.

Chapter 7
The Air Transport Insurance Industry

The air transport industry provides a sound analogy and comparison to the space insurance industry. Insurance coverage in the air transport industry carries the same objective as space insurance in that risk management in the overarching purpose of the insurance contract. A risk entails four possible responses from the person at risk: acceptance; elimination; reduction; and transfer.[1] The risk management aspect of insurance relates to the last element – transfer – whereby a person at risk would transfer the consequences of that risk to an insurer at a premium. The risks so transferred through insurance, particularly in relation to air transport, apply to the risks of theft, loss, or damage in a physical sense; bankruptcy, economic recession, decline or loss in a commercial sense; war, hijacking or repossession of aviation property in a political sense; natural disasters in an environmental sense; human resource problems in a social sense; business interruption in a financial sense; and legislative changes in a regulatory sense.[2]

The current crisis in risk management, particularly in transferring the risk of possible loss occurring to and through commercial air transport, is a direct corollary to the events of 11 September 2001. The international insurance market gave notice on 17 September that, effective 24 September, third party war risk liability insurance, covering airline operators and other service providers against losses and damages resulting from war, hijacking and other perils, would be cancelled.[3] The most compelling reason for the cancellations was the emergence of an exposure in terms of third party bodily injury and property damage that was unquantifiable. The International Union of Aviation Underwriters (IUAU) assessed that the total losses in respect of third party bodily injury and property damage caused by these events could exceed the previous greatest single catastrophic loss of US$20 billion caused by Hurricane Andrew in 1992 by a significant margin.[4]

[1]See Margo (1992) at p. 80.

[2]Margo (1992), pp. 79–80.

[3]Underwriters subsequently reinstated partial coverage for war risks with drastically reduced limits at considerably higher premia.

[4]See SGWA/1-IP/4, 5/12/01, Comments of IUAU, ICAO Special Group on Aviation War Risk (SGW1), First Meeting, Montreal, 6 to 7 December 2001 at p. 2.

R. Abeyratne, *Space Security Law*,
DOI 10.1007/978-3-642-16702-7_7,

As an immediate response to this measure, the President of the ICAO Council, Dr. Assad Kotaite, issued a State Letter[5] to all ICAO Contracting States, requesting that they take effective measures to preclude aviation and air transport services from coming to a standstill. This letter also appealed to Contracting States to support airline operators and other relevant parties, at least until the insurance market stabilized, by committing themselves to cover any risks to which airline operators and others may become exposed by the cancellation of insurance cover.

The 33rd Session of the ICAO Assembly, held in Montreal from 25 September to 5 October 2001, considered as an urgent priority the insurance issue by adopting Resolution A33-20.[6] This Resolution, while recognizing that the tragic events of 11 September had adversely affected the operations of airline operators globally as a result of war risk insurance cover no longer being available at levels which are practical and accessible to airline operators, *prima facie* urges Contracting States to work together to develop a more enduring and coordinated approach to the important problem of providing assistance to airline operators and other service providers. The Resolution, basing itself on the fundamental premise enunciated in Article 44 of the Chicago Convention,[7] which refers to the objective of ICAO to ensure safe, regular, efficient and economical air transport, directed the Council of ICAO to urgently establish a Special Group to consider issues emerging from action taken in the insurance market regarding third party war risk insurance coverage.

One must of course appreciate that war and associated risks, including hijacking and acts of terrorism, pose an extremely high risk exposure to insurers. Aviation hull and liability policies therefore usually contain an express exclusion in respect of such risks. The war risk exclusion used in the London market, known as AVN 48B,[8] excludes the risks of war, invasion, hostilities, civil war, rebellion, revolution, insurrection, martial law, hostile detonation of atomic weapons, strikes, riots, civil commotions or labour disturbances, acts of a political or terrorist nature, sabotage, confiscation, nationalization, seizure, and hijacking.

In practical terms, war risk insurance is required to cover three eventualities: to protect an airline operator from potential financial liability that could jeopardise its existence; to justify operations into territories of States by appeasing those States

[5]State Letter EC 2/6-01 dated 21 September 2001. The President of the Council followed this letter with two more letters, dated 25 October 2001 and 14 December 2001 respectively, appealing to all Contracting States to cover the risks left open until the insurance markets stabilized. The last letter also appealed to all Contracting States to extend or provide such coverage, as the case may be, until an international mechanism were put in place, thereby contributing to the stabilization of the markets.

[6]A33-20, Coordinated Approach in Providing Assistance in the Field of War Risk Insurance. See Assembly Resolutions in Force, (as of 5 October 2001), Doc 9790, ICAO: Montreal at V-6.

[7]*Convention on International Civil Aviation*, signed at Chicago on 7 December 1944. See ICAO Doc 7300/9, Ninth Edition, 2006.

[8]The London insurance market introduced the AVN 48B Clause after the Israeli raid on Beirut Airport on 28 December 1968. This war and hijacking risk exclusion clause is now included in every aviation hull and liability policy. This clause covers a wide range of eventualities including damage caused as a result of any malicious act or act of sabotage.

that they and their citizens would be financially compensated in the event of damage; and to protect the financial interests of airlines, their owners, financiers or lessors. It is usual for an aircraft, depending on its type, to be covered for any amount up to US$750 million to US$1 billion on aggregate (as against per single occurrence). As against this figure, it is significant that the underwriters permitted coverage for only up o US$50 million aggregate consequent upon their issuing notice of withdrawal of third party war risk insurance on 17 September 2001.

Many Contracting States, following the State Letter of the President of the ICAO Council, stepped in to address issues regarding cancellation of insurance. In the light of the dramatic recession of insurance coverage, States began to take measures to provide excess insurance cover to carriers, in most cases up to previous policy limit, for war and terrorism related third party risk. Provision of such coverage meant that at least some air carriers would not be in violation of domestic and international regulations and lease covenants respecting war risk cover. However, there was concern expressed with the fact that a considerable number of countries in Latin America, Asia and Africa, while having taken steps necessary to ensure continued coverage, have not provided the necessary guarantees and indemnities in the same amount as States in Europe and North America.

Action taken by ICAO Contracting States in responding to the insurance crisis has legal legitimacy in two international Conventions, i.e., the Rome Convention of 1952[9] and the Montreal Convention of 1999.[10] Article 15 of the Rome Convention provides that any Contracting State may require that the operator of an aircraft registered in another Contracting State shall be insured in respect of his liability for damage sustained in its territory for which a right to compensation exists. The operative clause, in the context of indemnities offered by the several ICAO Contracting States as discussed earlier, is contained in Article 15.4 of the Rome Convention which provides that, instead of insurance, *inter alia*, a guarantee given by the Contracting State where the aircraft is registered, shall be deemed satisfactory if that State undertakes that it will not claim immunity from suit in respect of that guarantee.[11] The Montreal Convention of 1999 provides in Article 50 that States Parties shall require their carriers to maintain adequate insurance covering their liability under the Convention. This provision further stipulates that a carrier may be required by the State Party into which it operates to furnish evidence that it maintains adequate insurance covering its liability under the Convention.

It must be noted that coverage provided by airline insurance policies regarding perils other than third party liability for war risks have not been affected by this cancellation. War and allied perils coverage with regard to passengers have been left unchanged but the uncertainty created by the events have made it essential to

[9]Convention on Damage Caused by Foreign Aircraft to Third Parties on the Surface, signed at Rome on 7 October 1952, ICAO Doc 7364.

[10]Convention for the Unification of Certain Rules for International Carriage by Air, done at Montreal on 28 May 1999, ICAO Doc 9740.

[11]See Article 15.4.(c).

circumscribe coverage for third party losses at a maximum of US$50 million. Although, increasing due to a sustained period of unprofitable trading in the insurance market, the events themselves triggered accelerated premium increases both in order to assist markets to revive from the bout of unprofitable trading and to create a reasonably adequate premium base for future exigencies of the nature of the catastrophes of September 2001.

In general terms, the price to be paid to revive or reinstate adequate coverage for third party was risk coverage would cost the airlines an additional premium of US $1.25 per passenger carried. If airlines were to purchase coverage for limits of US $950 million in excess of the already available US$50 million they would have to pay US$1.85 per passenger carried. In view of the fact the airports, refuellers, ground handlers and other service providers in the aviation industry contribute to an accumulation of risk, since many of them may serve a particular airline at one location, underwriters were disinclined to offer coverage for these providers. However, many insurers have shown willingness to extend coverage for an additional US$100 million over the US$50 million coverage already provided.[12]

Both the International Civil Aviation Organization (ICAO) and the International Air Transport Association (IATA) have stringently and correctly maintained that there is an inherent role to be played by governments in the event of war risk claims. IATA has justifiably claimed, in a well reasoned argument,[13] that a new international regime must provide for governments to agree to act as a multilateral guarantor covering terrorist actions against airlines in any part of the world. IATA has requested that any solution to the insurance crisis be widely available to international aviation shareholders; be reasonably affordable; provide for long term stability even in the event of terrorist acts; and recognize the inherent role of governments in the event of war risk claims.

The above remarks were made at the First Meeting of the ICAO Special Group on War Risk Insurance, held in Montreal on 6–7 December 2001. This special group was appointed by the ICAO Council in response to ICAO Assembly Resolution A33-20,[14] adopted at the 33rd Session of the Assembly in September/October 2001. As earlier stated, this resolution urges Contracting States to cooperate in developing a non-enduring coordinated approach to the important problem of providing assistance to airline operators and to other service providers in the field of aviation war risk insurance. Toward achieving this objective, the Assembly directed the Council to urgently establish a Special Group to consider the issues referred to above and to report back to Council with recommendations as soon as possible. The resolution also broadens ICAO's mandate by inviting the Council and the Secretary General to take any other measures considered necessary or desirable.

---

[12]See Article 15.4.(c).

[13]SGW1/1-WP/4, 5/12/01 – Comments of IATA, Special Group on Aviation War Risk Insurance, First Meeting, Montreal, 6 to 7 December 2001 at p. 2.

[14]Resolution A33-22 – *Coordinated approach in providing assistance in the field of war risk insurance*, Assembly Resolutions in Force (as of 28 September 2007) ICAO Doc 9902, at V-9.

At the second meeting of the Special Group, held in Montreal from 18 to 20 January 2002, The London Market Brokers Committee (LMBC) presented a medium term scheme to cover airlines from war risk liabilities. The scheme envisions the formation of a company, the board of directors of which shall include representatives of participating States, ICAO and participating aviation and insurance industries. The company would offer third party war risk liability cover up to US$1.5 billion in excess of US$50 million per insured. This cover will be non-cancellable, apply per occurrence and per aircraft where multiple aircraft are involved. The insurance cover to be provided by the company would be available to the entire aviation sector and include domestic and international operations as well as equipment lessors, financiers and manufacturers of each State that joins the scheme.

The scheme so outlined offers a continuous cover of aviation war and other perils liability insurance based on clauses AVN52D and AVN52F (which generally exclude coverage of war risk liability with a write back possibility). The scheme also admits of a full review by participating Contracting States, to be undertaken at its fifth anniversary, with an option to cancel/suspend the scheme 90 days thereafter.

Participating States would act as guarantors or "reinsurers of last resort" through a legal agreement with the insurance company. In the event of a claim, the contributions of participating States would be pro-rated based on their ICAO assessments. Each State's maximum liability under the scheme would be capped. The total cap, if all ICAO States participate in the scheme, is expected to be US$15 billion (therefore, for example, if only 50% of ICAO Contracting States participate, the total cap would be US$7.5 billion). The maximum exposure of each State, in any given instance, would be its ICAO assessment percentage of the total cap as it may apply, depending on the participation of States in the scheme, as outlined above.

Premiums will be collected from the insured in order to build a reinsurance pool to meet claims under the policies. This pool will obviate the need for participating States to make cash contributions to the company in the event of a claim. The total amount of premiums to be collected in the first year is targeted at US$850 million (equivalent to 50 cents per passenger segment based on total passenger segments of 1.7 billion). The premiums for subsequent years would be kept at approximately the same level, provided there were no losses.

Although, some members argued that the US$0.50 per passenger charge was not an equitable measurement for the collection of the premium (as the numbers carried per flight may differ and smaller aircraft may not necessarily be considered as much a threat as weapons of destruction as the larger aircraft which have larger capacity), the Group decided to work on the basis of US$0.50 per passenger as this was considered to be the only workable means of premium funding.

The work of the Special Group was considered by a Council Study Group on Aviation War Risk Insurance, established by agreement of the Council on 4 March 2002. This Study Group had two meetings, on 16 April and 24 April 2002 respectively, wherein the Group considered a draft report to Council containing the outcome of the work of the Special Group. This report firstly outlines coverage to be provided in respect of third party war risk liability insurance, which is up to

US$1.5 billion per aircraft, per occurrence, per insured, over and above the coverage offered by the private market amounting to US$50 million, which is already in place. Special features, which tantamount to advantages offered by this coverage are that it would not be cancellable (which is in contrast to the current 7 day cancellation clause) and that coverage would encompass all areas of the aviation industry, including airlines, airports, ground handling agents, screening companies, manufacturers of aircraft and components lessors, air traffic controllers and other providers of air navigation services. The scope of coverage would be global.

In terms of rates, the ICAO scheme would charge 50 cents per passenger for coverage up to US$1.5 billion in excess of the private cover of US$50 million, which, as already mentioned, is available at US$1.25 per passenger. The rate of 50 cents per passenger compares favourably with the current US$1.50 excess charge currently levied in respect of excess third party insurance which goes only upto a maximum of US$1 billion in two extra layers at US$1.00 for both layers in addition to the primary cover fixed at US$1.25. The premium advantage notwithstanding, the strongest thrust of the coverage offered by the ICAO scheme, however, remains in its intrinsic guarantee against cancellation, particularly in view of the existing 7-day cancellation clause.

With regard to participation, which is of course voluntary, the exposure of a participating State to risk of payment in the instance of a claim under third party war risk liability would amount to its ICAO contribution percentage of US$1.5 billion. To give an example, a State which participates in the ICAO scheme which contributes 3% of the ICAO budget has a maximum exposure of US$45 million. Compared to State guarantees given in the aftermath of the September 2001 events, which were often unlimited, this modality should be acceptable to most States. In order to participate, an ICAO contracting State would be required to sign a participation agreement with ICAO, which would be generally designed to fit the particular legal structure and legislative requirements of each State concerned.

An insurance entity which is proposed within the parameters of the ICAO scheme would have to be established by the ICAO Council, and thereafter be formally incorporated jointly by ICAO and the industry, consequent upon development of appropriate statutes and statutory instruments, in accordance with applicable domestic and regulatory requirements. The participation agreement would be open for signature to all ICAO Contracting States.

At the Third Meeting of the Council Study Group on Aviation War Risk Insurance, held at ICAO on 14 January 2003, the Study Group considered the status of developments since its second meeting, noting that 45 Contracting States had indicated their intent to participate in the global war risk insurance scheme whereas 10 States had responded negatively, expressing their unwillingness to participate.[15] The Group also considered a revised Draft Participation Agreement for the Global Scheme Regarding the Provision of Aviation War Risk Insurance[16] (which had been circulated earlier to contracting States).[17] This draft Agreement is designed to

[15]CGW1/3-IP/1. 18 States had reserved their position with regard to participation.

establish an Insurance Entity (IE) for the sole purpose of providing aviation insurance cover on prescribed terms for war and allied perils related liability risks faced by airline operators and other commercial entities providing aviation related services. The purpose of the agreement which is mainly to obtain from participating States a guarantee certain obligations of th IE and to establish the proration, limits and payment mechanisms related such obligations – in other words to provide complimentary cover through the IE that was withdrawn or reduced by the commercial insurance market following the events of 11 September 2001.

With regard to the scope of coverage, the IE will provide aviation war risk cover from the excess point per insured up to US$1.5 billion. The same amount would apply to operators who have cover under AVN 52D and AVN 52F clauses or any derivatives thereof, on the basis the amount would apply to any one occurrence, any one aircraft and any one insured. This limit of $1.5 billion will be applicable in addition to the primary passenger and third party limits that were provided by the insurance markets prior to 11 September 2001. A lower limit is also provided under the IE coverage of $500 million for operators who obtain coverage under AVN 52E and AVN 52G or derivatives of such coverage. The IE's cover shall automatically apply to those who are originally insured and who lose their war risk coverage as dictated by the insurance market (when such third party cover is up to the excess point or passenger war risk insurance cover) under their primary aviation insurance policies. insurance In the case of passenger war risk cover, the limits of 1.5 billion will be raised to $2 billion and up to $750 million respectively.

The IE will, under the participation agreement, meet any claims through funds accumulated from premiums, earned investment income and income from other sources, along with borrowings, while participating States will remain as guarantors of last resort. Premiums will be collected from original insureds who are air carriers designated for the purpose of the Agreement by State parties; any lessors, financiers and manufacturers incorporated in a participating State (State Party) who purchase their own primary insurance; and any service provider incorporated in a participating State who is in the business of providing services or goods in that State to any person or entity engaged in the aviation industry. Any other person or entity identified by the above categories of original insureds as additional insureds would be exempt from payment of premiums. The Agreement make provision for the IE to seek borrowings from credit institutions in the event funds accumulated through financial resources identified above are not sufficient to meet claims. The IE is required to maintain at all times liability insurance covering the interests of directors, officers and employees of the Entity.

For the part of participating States, their obligations are to guarantee to the IE that they will meet claims arising from insurance policies issued by the IE to

[16]CGWW1/3-WP1, Appendix A.

[17]See Attachment B to State Letter LE 4/64-02/55.

original insureds incorporated in the territory of a signatory State or any other participating State Party to the agreement. The participating States also warrant that the agreement would, for all purposes, be treated as a commercial agreement, i.e., a contract.

The inherent advantages of the proposed ICAO scheme are its uniqueness in terms of its global application, non cancel ability, afford ability with regard to premium and exposure to claims, and its design in accordance with regulatory requirements.

The commonality between the problems of space insurance and air transport insurance lies in the enormity of exposure to risk faced by both industries. In the context of space insurance, underwriters are primarily concerned with the rapidity with which the sequence of spacecraft failures occur. One of the reasons for satellite failure may well be the accelerated rate of their manufacture, which has shortened from 36 months to 12 months. The reliance on generic spacecraft specifications could also be a contributory factor. With regard to commercial air transport insurance, the increased exposure to risk is particularly in the field of security and the threat of unlawful interference. Additionally, the safety of aviation is also a concern, sometimes conceptually attributed to the proliferation of flights by carriers to attain commercial expediency and provide for an increasing demand for air services. Whatever may be the reasons for increased exposure to risk, both space and air transportation must, of necessity, address the compelling need to review ways and means of ensuring adequate provision of insurance coverage.

One of the issues that would be relevant, and be politically and socially compelling, is the extent to which States can be called upon to be responsible for ensuring that both these critical areas are covered for risks so that continuity of the services they render are assured. The reason for this is clear. The space and insurance industries clearly suffered a paradigm shift, largely brought to bear by the impact of the events of 11 September 2001 on the air transport industry. Before these events, war risk insurance coverage for aviation, which was included in the standard insurance policy, was the obscure preoccupation of insurance managers of airlines. The only "red flag" in the war risk coverage was the 7-day cancellation clause which was seldom invoked until the events of September 2001.

The fluctuating and untenable situation in the space and air transport insurance industries demonstrate that both industries are "brittle" and therefore susceptible to catalysts of market failure. It is this reason that calls for States to be insurers of first resort rather than last resort. States should play the role of initiator and regulator of insurance to the extent of ensuring that insurance is available rather than actually providing it. The ultimate provision of insurance should be left to the commercial insurance market.

The synergies between air transport and space insurance are seen particularly in war risk insurance, where substantial neglect on the part of a State to take reasonable preventive or preemptive action, and neglect due to lack of attention, official indifference or connivance will impose upon that State responsibility for damage to foreign, private and public property.[18] Such a responsibility could give rise to the legal remedy of *restitutio in integrum*, usually granted to the injured person by a tribunal by way of a declaration, or by restitution in kind or specific restitution. Additionally, the rule of law requires that, if damage is caused by negligence in the course of a lawful activity, the award of compensation may be a lega remedy.[19] This is one more reason for States to be interested in involvement one way or another in the regulatory process or guidance-setting with regard to insurance coverage.

A tangible example and experience has already been provided by ICAO in its offer to the aviation community of a viable regulatory process with regard to air transport insurance. ICAO's involvement, until 17 September 2001 when the underwriters gave 7 days' notice of withdrawal of war risk coverage, was nonexistent. Now, its role is predominant, at the request of the 188 Contracting States of the Organization.

The most fundamental commonality in the paradigm outlining the purpose of both the areas of outer space and air transport activity is essentially that both are for the benefit of the public good and the well being of nations and therefore, any suspension of activity would be seriously detrimental to the welfare of common humanity. In the outer space regime, the benefits accorded by space exploration to both States and people is explicitly recognized in Article 1 of the *Outer Space Treaty* which provides that the exploration and use of outer space, including the Moon and other celestial bodies, shall be carried out for the benefit and interests of all countries, irrespective of their economic or scientific development, and shall be the province of all mankind. In the air transport field, the *Convention on International Civil Aviation*,[20] in its *Preamble*, recognizes that whereas the future development of international civil aviation can greatly help to create and preserve friendship and understanding among nations and peoples of the world (thus recognizing, as in the context of outer space activity, that both States and people benefit from air transport), yet its abuse (i.e., abuse of the future development of civil aviation) can become a threat to the general security. One can find no compelling legal pronouncements stronger than these to conclude that States are necessarily and integrally involved in assuring the sustainability of outer space activities and air transportation.

The final issue to be addressed, in terms of State involvement as the first insurer or insurer of first resort, is the manner in which State involvement can be consolidated. It must, as of necessity be through international treaty where consent of the States' parties to be bound by such a treaty will be a legal prerequisite. It is only in this manner that insurance at last resort can be ensured through the commercial insurance market at reasonable rates. Preference for one over the other, as is

---

[18]See the Younans Case, (1926) RIAA iv 110 at 116; 21 AJ (1927) 571 at p. 578. See also U.K. and Indonesia, exchange of notes, 1 December 1966, Treaty series, no. 34 (1967).

[19]See Brownlie (1992) at pp. 462–464.

[20]Global Positioning System Report published by Ireland based Research and Markets, Online information available at http://www.researchandmarkets.com/reports/3687/3687.htm (Accessed on 12 August 2006).

currently occurring in the air transport and space industry, can then be effectively precluded.

References

Brownlie I (1992) Principles of public international law, 4th edn. Clarendon Press, Oxford, pp 462–464

Margo RD (1992) Risk management and insurance. Ann Air Space Law 17(1):79

Chapter 8
The Application of Intellectual Property Rights to Outer Space Activities

Some years ago, an informal assessment revealed that 3–6% of world trade is carried out with counterfeit and pirated goods, which in real terms amounts to approximately US$120–240 billion per annum.[1] The possibility is very real that this figure is now even higher. Although there is no known assessment of this kind pertaining to trade in space equipment, it is relevant and necessary to address the issue of intellectual property rights and their acquisition with regard to outer space activities, particularly pertaining to the proliferation of research activity that is now being carried out by space-faring nations.

The acquisition of intellectual property rights is accomplished intrinsically on a territorial basis. Outer space activity is essentially extra territorial in that a State engaging in outer space activities cannot claim territoriality in outer space for such activities. This dichotomy seemingly sets the stage for an inconsistency in the application of intellectual property laws to objects and activities in outer space. However, this perception is clearly *ex facie* illusionary, since outer space activity and objects used in outer space start for the most part on Earth. For instance, a lunar module or "moon buggy" invented and manufactured on Earth may clearly be drawn into the regime of intellectual property laws of a State concerned. However, the difficulty would arise if a State, in the course of its outer space activities conducted extra terrestrially, produces in space an object or machine which would not have a territorial link since there is no room for acquisitioning property rights in outer space. This chapter will examine the principles applying to the acquisition of intellectual property rights, if any, under circumstances linking outer space activity.

Territoriality

Intellectual property rights can be acquired and applied in two ways: territorially and internationally. For instance, if an invention is registered in Canada, the rights accruing to the person registering that invention's patent in Canada applies only

[1] Vandoran (1999), p. 25 at 26.

DOI 10.1007/978-3-642-16702-7_8,

within that country. Such a right cannot be infringed by acts perpetrated in the United States. An aggrieved person whose intellectual property right is infringed can only seek redress against the injury in the country in which his right was infringed, according to the laws of that country. However, intellectual property rights are also applicable internationally and their existence will not be restricted to the jurisdiction of the State in which the activity creating such rights took place. Thus an invention in State A can be patented in State B and a literary work created in State C may acquire copyright in State D automatically. A complex web of international treaties protect intellectual property rights of holders on a transboundary basis, primarily to obviate discrimination against foreign patent, trademark or copyright owners in a local jurisdiction. A good example is Canadian Law which has succeeded in harmonizing equitable application of intellectual property laws both nationally and internationally, as applicable to an instance of adjudication in Canada.[2]

The national or territorial concept of intellectual property rights creates a dichotomy where it clashes with the transboundary or international application of rights. This clash may occur particularly in the field of communications technology. For example, if State A were to download certain material and data pertaining to an outer space project in which it is involved, and it is picked up by another State and transmitted to its space station in outer space, or more seriously, if a space station of a country other than State A were to directly access and use such material and data, exclusively in outer space, would State A have any recourse to terrestrial or territorial intellectual property laws against such usage?

Arguably, the strongest proposition supporting the application of intellectual property laws to outer space activities would lie in a contrived process of reasoning, starting with the fundamental premise that the Outer Space Treaty of 1967[3] which lays down the principle that no State can claim sovereignty over any portion of outer space.[4] Since the concept of sovereignty connotes ineluctably a territorial control by that State, the Outer Space Treaty effectively precludes a State from exercising this right in outer space. However, this does not necessarily mean that a State has no right or control over its space objects or space personnel in outer space. Article VI of the Outer Space Treaty ensures that States have the right to require authorization and continued supervision by that State on the activities in outer space by a non Governmental organization or entity of that State. Furthermore, Article VIII of the Treaty provides for a State to retain jurisdiction and control over an

[2]See *National Corn Growers Association* v. *Canada* (Import Tribunal) [1990] 2.S.C.R 1324, 74 D. L.R (4th) 449 at 482-83 (S.C.C) (GATT). Also, *Milliken & Co.* v. *Interface Flooring Systems (Canada) Inc.* (1993) 52.C.P.R. (3d) 92 (Fed.T.D.) affirmed (1994), 58 C.P.R (3d) 157 (Fed.C.A).

[3]Treaty on Principles Governing the Activities of States in the Exploration and Use of Outer Space, Including the Moon and other Celestial Bodies, January 27, 1967 (entered into force on 10 October 1967), 610 UNTS 205.

[4]Treaty on Principles Governing the Activities of States in the Exploration and Use of Outer Space, Including the Moon and other Celestial Bodies, January 27, 1967 (entered into force on 10 October 1967), 610 UNTS 205 Article 1.

object launched into outer space and over any personnel, if such object were registered in that State. *A fortiori*, the Inter Governmental Agreement of 1998 Relating to the International Space Station has an explicit provision that ensures territorial application to objects registered in the State Parties[5] concerned. Thus, a module belonging to a particular State Party to the Inter Governmental Agreement or is registered in that State, and is an integral component of the space station, would be deemed "territory" of the State concerned. This principle is embodied in Article 21(2) of the Agreement which provides that any invention made on a module of a space station will be deemed to have been made in the State to which that module belongs, and that any activity occurring in or on a space station flight element shall be deemed to have occurred exclusively in the territory of a State in which that element is registered.

Space Law Principles

The legal and philosophical bases of space law are unique and form the antithesis of those applicable to air law in that space law is grounded on the principle that outer space is the common heritage of mankind and that no State or individual can therefore claim *rights in rem* to any portion of outer space. Air law, on the other hand, is firmly entrenched in the principle of sovereignty of States, so that a State may lay claims to rights over the airspace above its territory.[6] Thus in aviation, general principles applicable to intellectual property rights would apply. This essentially means that while the implementation of air law is heavily influenced by municipal law, space law is solely grounded on legal principles binding on the community of nations. Principles of public international law therefore play an exclusive part in the application of space law principles.

In terms of jurisprudence, space law represents the Idealist school which supports community interest over national interest, while air law represents the Realist school which advocates that national interests are pre-eminent considerations for all purposes. The Idealist school believes that individual interests should best be served by collective intercourse and is best illustrated by the view of Lauterpacht who was of the view:

> a community may be said to be the body of a number of individuals more or less bound together through such common interests as to create a manifold intercourse between single individuals.[7]

---

[5]State Parties are Canada, member States of the European Space Agency, Japan, Russian Federation and United States of America.

[6]See Abeyratne (1992), pp. 135–144.

[7]Lauterpacht (1955), p. 11.

Legal principles relating to the international community necessarily emanate collectively from that community as a body of rules which require the consent of the community. An examination of the philosophy of space law therefore essentially requires an examination of the nature of public international law itself. This chapter will discuss the philosophical basis of the common heritage principle of space law, through an evaluation of public international law and its relation to each other.

Space law is one of the most recent additions to international jurisprudence. That it pertains to one of the most highly technology intensive activities is an incontrovertible fact and was made evident with the successful launch of the Space Shuttle "Columbia" on 12 April 1981, where the world entered a new age of space exploitation, leaving behind the period of space exploration which seemingly started in 1957 with the launch of the Russian "Sputnik."

The emergent philosophical problem posed by space law, in its offer to mankind of a new dimension of transportation law and property law, was succinctly subsumed by Professor Bockstiegel in 1983:

> (Space law) ... is the newest main field of international law ... and it depends more than most other fields on probable and fast technical progress ...[8] It is obvious that the application of space technology will lead to the growing commercialization of space activities, since such service - at least in the long run - can only be maintained and expanded, if it is self financing ...[9]

The blending of high technology with a new forensic code of conduct on hitherto unchartered territory has brought to bear the need for the community of nations to formulate a sustainable legal theory that would ensure non-exploitation of space resources by individuals or States, while at the same time incorporating the element of responsibility and liability for individual and State conduct in outer space.

As mentioned earlier, the basic principle of space law is the "common interest" (or common heritage) principle which emerged as a result of the first specific Resolution on space law of the United Nations General Assembly in 1958.[10] The "common interest" principle has since been incorporated in subsequent multilateral treaties, particularly the *Outer Space Treaty* of 1967,[11] Article 1(1) which provides:

> [T]he exploration and use of outer space, including the moon and other celestial bodies, shall be carried out for the benefit and in the interest of all countries, irrespective of their degree of economic or scientific development, and shall be the province of all mankind.

This provision, which binds signatory States, is seemingly a departure from the traditional "national interest" approach of international air law and has represented

---

[8]Böckstiegel (1983), p. 305.

[9]Böckstiegel (1983), at p. 314.

[10]*UNGA Resolution 1348 (XII)*, 13 December 1958.

[11]*Treaty on Principles Governing the Activities of States in the Exploration and Use of Outer Space, Including the Moon and Other Celestial Bodies*, opened for Signature at Moscow, London and Washington on 27 January 1967, 610 UNTS 205.

a moral obligation to some,[12] while to others the provision has represented a *jus cogens* or mandatory legal principle.[13]

The International Court of Justice (ICJ), in the *North Sea Continental Shelf Case,*[14] held that legal principles that are incorporated in Treaties, such as the "common interest" principle, become customary international law by virtue of Article 38 of the 1969 Vienna Convention on the Law of Treaties. Article 38 recognizes that a rule set forth in a treaty would become binding upon a third State as a customary rule of international law if it is generally recognized by the States concerned as such. Article 1(1) of the *Outer Space Treaty*, which designates that the use of space technology is achieved under the "common interest" principle for the common good of humanity, therefore becomes a principle of customary international law, or *jus cogens*. Obligations arising from *jus cogens* are considered applicable *erga omnes* which would mean that States using space technology owe a duty of care to the world at large in their outer space explorations.

In this regard:

> provisions are found on the fundamental moral principles contained in the foregoing paragraphs of Article 1 and 2 of the UN Charter, such basic principles qualify for the status of peremptory norms of general international law.[15]

The effect of this observation is that the content and nature of Article 1 (1) confirms that it is a *jus cogens*. There is seemingly no reason why the international community should not give such recognition to the "common interest" principle as enshrined in Article 1(1) which is aimed at the protection of the interests of the international community as a whole. *A fortiori*, on the same basis, Article IX of the *Outer Space Treaty* which requires that States should avoid harmful contamination and adverse change in the environment of the Earth which may result from the exploration of outer space would incontrovertibly be considered *jus cogens*.

As was stated earlier, Article VI of the *Outer Space Treaty* apportions responsibility on States who are Parties to the Treaty for national activities in outer space, whether such activities are carried out by governmental agencies or non-governmental agencies. This provision clearly introduces the notion of strict liability *erga omnes* to the application of the *jus cogens* principle relating to outer space activities of States and could be considered applicable in instances where States hold out to the international community as providers of technology achieved and used by them in outer space, which is used for purposes of air the provision of such technology. The ICJ in the *Barcelona Traction Case* held:

> [A]n essential distinction should be drawn between the obligations of a State towards the international community as a whole, and those arising *vis à vis* another State in the field of diplomatic protection. By their very nature, the former are the concerns of all States.

---

[12]Goedhuis (1976), p. 195 at pp. 198–199. Also Cheng (1968), p. 532 at 578.

[13]Markoff (1976), p. 3. Also, Matte (1980), p. 119 at 147. Jakhu, Christol (1983), p. 1.

[14]*I.C.J. Reports* 1970, at 32.

[15]Christol (1983) at 6.

In view of the importance of the rights involved, all States can be held to have a legal interest in their protection; they are obligations *erga omnes*.[16]

The International Law Commission has observed of the ICJ decision:

[I]n the Courts view, there are in fact a number, albeit limited, of international obligations which, by reason of their importance to the international community as a whole, are- unlike others - obligations in respect of which all States have legal interest.[17]

The views of the ICJ and the International Law Commission, which has supported the approach taken by the ICJ, give rise to two possible conclusions relating to *jus cogens* and its resultant obligations *erga omnes*:

(a) obligations *erga omnes* affect all States and thus cannot be made inapplicable to a State or group of States by an exclusive clause in a treaty or other document reflecting legal obligations without the consent of the international community as a whole; and
(b) obligations *erga omnes* preempt other obligations which may be incompatible with them.

Some examples of obligations *erga omnes* cited by the ICJ are prohibition of acts of aggression, genocide, slavery and discrimination.[18] It is indeed worthy of note that all these obligations are derivatives of norms which are *jus cogens* at international law.

If it can be accepted that a principle of *jus cogens* creates obligations *erga omnes*, it becomes an undeniable fact that Article 1(1) of the *Outer Space Treaty* could be considered a peremptory norm or *jus cogens,* since it generates obligations towards the international community as a whole. Christol observes:

Article 1 para. 1 of the Space Treaty, with its adoption of the common benefits and interests guarantee, can be supported (as an example of peremptory norms) because the provisions conform to moral law in the sense that all humankind is to benefit unconditionally, and because the terms are consistent with the spirit and the purposes identified in Article 1 paras. 1 through 3 and Article 2 paras. 1 through 4 of the UN Charter, as well as with complimentary international agreements of lesser authority. To the extent that the terms are beneficial to individuals, the larger community, and States, and when the navigation. Article VI further requires that the activities of non-governmental entities in outer space shall require authorization and continuing supervision by the appropriate State Party to the Treaty, thus ensuring that the State whose nationality the entity bears would be vicariously answerable for the activities of that organization, thereby imputing liability to the State concerned. Article VII makes a State Party internationally liable to another State Party for damage caused by a space object launched by that State.

[16] *Barcelona Traction, Light and Power Company Limited, I.C.J. Reports, 1974*, p. 253 at pp. 269–270.

[17] *Yearbook of International Law Commission* 1976, Vol II, Part One at 29.

[18] *I.C.J. Reports*, 1970 at 32.

The *Registration Convention* of 1974[19] in Article II(1) requires a launching State of a space object that is launched into earth orbit or beyond, to register such space object by means of an entry in an appropriate registry which it shall maintains and inform the Secretary General of the United Nations of the establishment of such a registry. This provision ensures that the international community is kept aware of which State is responsible for which space object and enables the United Nations to observe outer space activities of States. Article VI of the Convention makes it an obligation of all State Parties, including those that possess space monitoring and tracking facilities, to render assistance in identifying a space object which causes damage to other space objects or persons. Justice Manfred Lachs analyses these provisions of the *Registration Convention* to mean that the State of registry and the location of the space object would govern jurisdictional issues arising out of the legal status of space objects.[20] On the issue of joint launching of space objects, Justice Lachs observes:

> No difficulties arise whenever a State launches its own object from its own territory; the same applies to objects owned or launched by non-governmental agencies registered in that State. However, in cases of joint launching, agreement between the parties is required as to which of them is to be deemed the 'State of Registry'. A similar agreement is also necessary when a launching is carried out by an international organization.[21]

The above provision ensures the identification of parties responsible for specific activities in outer space and thereby makes it easier to impose liability for environmental damage caused.

The *Outer Space Treaty*,[22] while expostulating the fundamental principle in its Article 1 that the exploration and use of outer space, including the moon and other celestial bodies, shall be carried out for the benefit and in the interests of all countries, explicitly imposes in Article VII international liability and responsibility on each State Party to the Treaty, for damage caused to another State Party or to its populace (whether national or juridical) by the launch or procurement of launch of an object into outer space. In its preceding provisions the Treaty imposes international responsibility on States Parties for national activities conducted in outer space. The Treaty also requires its States Parties to be guided by the principle of co-operation and mutual assistance in the conduct of all their activities in outer space.[23] This overall principle is further elucidated in the same provision:

---

[19]*Convention on Registration of Objects Launched into Outer Space*, Adopted by the General Assembly of the United Nations, New York, 12 November 1974, 1023 UNTS 15.

[20]Lachs (1972), p. 70.

[21]Lachs (1972), p. 70.

[22]*Treaty on Principles Governing the Activities of States in the Exploration and Use of Outer Space, including the Moon and Other Celestial Bodies*, opened for signature at Moscow, London and Washington, 27 January 1967, 610 UNTS 205, http://www.telegraph.co.uk/science/space/7631252/Stephen-Hawking-alien-life-is-out-there-scientist-warns.html.

[23]*Treaty on Principles Governing the Activities of States in the Exploration and Use of Outer Space, including the Moon and Other Celestial Bodies*, opened for signature at Moscow, London

> States Parties to the Treaty shall pursue studies of outer space, including the moon and other celestial bodies, and conduct exploration of them so as to avoid harmful contamination and also adverse changes in the environment of the Earth resulting from the introduction of extra terrestrial matter.[24]

The *Moon Agreement*[25]of 1979 provides that in the exploration and use of the moon, States Parties shall take measures *inter alia* to avoid harmfully affecting the environment of the earth through the introduction of extra terrestrial matter or otherwise.[26]

The *Liability Convention*[27] contains a provision which lays down the legal remedy in instances of damage caused by Space objects. Article II provides:

> A launching State shall be absolutely liable to pay compensation for damage caused by its space objects on the surface of the Earth or to aircraft in flight.,[28]

thereby imposing a regime of absolute liability on the State that launches space objects such as satellites, which provide technology and communication that is used for air navigational purposes. Although admittedly, both the *Outer Space Treaty* and the *Liability Convention* do not explicitly provide for damage caused by technology and communication provided by space objects, culpability arising from the "common interest" principle and liability provisions of the two conventions can be imputed to States under these Conventions.

Gorove states that in the field of international space law, two clearly connected terms have been used: liability and responsibility.[29] Although "responsibility" has not been cohesively interpreted in any legal treaty relating to outer space, "liability" occurs in the *Liability Convention* and is sufficiently clear therein. This, however, does not mean that State responsibility is not relevant to the obligations of States law as, in international relations, the invasion of a right or other legal interest of one subject of the law by another inevitably creates legal responsibility. Professor Brownlie observes:

and Washington, 27 January 1967, 610 UNTS 205, http://www.telegraph.co.uk/science/space/7631252/Stephen-Hawking-alien-life-is-out-there-scientist-warns.html. Article IX.

[24]*Treaty on Principles Governing the Activities of States in the Exploration and Use of Outer Space, including the Moon and Other Celestial Bodies*, opened for signature at Moscow, London and Washington, 27 January 1967, 610 UNTS 205, http://www.telegraph.co.uk/science/space/7631252/Stephen-Hawking-alien-life-is-out-there-scientist-warns.html. Article IX.

[25]Agreement Governing the Activities of States on the Moon and other Celestial Bodies, signed on 5 December 1979, UN Doc A/RES/34/68 of 5/12/1979.

[26]Agreement Governing the Activities of States on the Moon and other Celestial Bodies, signed on 5 December 1979, UN Doc A/RES/34/68 of 5/12/1979. Article 7.

[27]*Convention on International Liability for Damage Caused by Space Objects*, March 29 1972, 24 U.S.T 2389, T.I.A.S No. 7762.

[28]Article II(a) defines damage as including loss of life, personal, injury or other impairment of health; or loss or damage to property of States or of persons natural or juridical, or property of international governmental organizations.

[29]Gorove (1983), p. 373.

> [T]oday, one can regard responsibility as a general principle of international law, a concomitant of substantive rules and of the supposition that acts and omissions may be categorized as illegal by reference to the rules establishing rights and duties. Shortly, the law of responsibility is concerned with the incidence and consequence of illegal acts, and particularly the payment of compensation for loss caused.[30]

International responsibility relates both to breaches of treaty provisions and other breaches of legal duty. In the *Spanish Zone of Morocco Claims* case, Justice Huber observed:

> [R]esponsibility is the necessary corollary of a right. All rights of an international character involve international responsibility. If the obligation in question is not met, responsibility entails the duty to make reparation.[31]

There is also explicit recognition that principles of international law apply to space law. The General Assembly of the United Nations in 1961 adopted the view that international law, including the Charter of the United Nations, applies to outer space and celestial bodies.[32] It is also now recognized as a principle of international law that the breach of a duty involves an obligation to make reparation appropriately and adequately. This reparation is regarded as the indispensable complement of a failure to apply a convention and is applied as an inarticulate premise that need not be stated in the breached convention itself.[33] The ICJ affirmed this principle in 1949 in the *Corfu Channel* Case[34] by holding that Albania was responsible under international law to pay compensation to the United Kingdom for not warning that Albania had laid mines in Albanian waters which caused explosions, damaging ships belonging to the United Kingdom. Since the treaty law provisions of liability and the general principles of international law as discussed complement each other in endorsing the liability of States to compensate for damage caused by space objects, there is no contention as to whether in the use of nuclear power sources in outer space, damage caused by the uses of space objects or use thereof would not go uncompensated. The rationale for the award of compensation is explicitly included in Article XII of the *Liability Convention* which requires that the person aggrieved or injured should be restored (by the award of compensation to him) to the condition in which he would have been if the damage had not occurred. Furthermore, under the principles of international law, moral damages based on pain, suffering and humiliation, as well as on other considerations, are considered recoverable.[35]

---

[30]Brownlie (1990), p. 433.

[31]*1925 RIAA ii* 615 at 641.

[32]Resolution 1721 (XVI) adopted on 20 December 1961. See also Article 3 of the *Outer Space Treaty*.

[33]*In Re. Chorzow Factory (Jurisdiction) Case, (1927) PCIJ, Ser. A, no. 9* at 21.

[34]*ICJ Reports (1949)*, 4 at 23.

[35]Christol (1991), p. 231.

As discussed, both treaty law and general principles of international law on the subject of space law make the two elements of liability and responsibility a means to an end - that of awarding compensation to an aggrieved State or other subject under the law. Therefore, in view of the many legal issues that may arise, the primary purpose of a regulatory body which sets standards on State liability in issues concerning the use of space technology would be to carefully consider the subtleties of responsibility and liability and explore their consequences on States and others involved as they apply to the overall concept of the status of a State as a user of space technology which may cause harm or injury to the latter.

Trade Related Activities of Intellectual Property

The transfer of technology and its symbiotic application, particularly in the sharing of technology among outer space faring nations of the world is critical to the progress of outer space activity. Therefore, as any other activity involving intellectual property rights, space law can attenuate for its basic principles the establishment of the World Intellectual Property Organization (WIPO) which was set up as a specialized agency of the United Nations in 1974.[36] The nature of the WIPO structure and the numerous conventions the Organization had to administer rendered its effects on intellectual property rights administration on a global scale somewhat ineffective. The inadequacy of the WIPO mechanism prompted industrialized nations to seek an alternative, which they found under the umbrella of the General Agreement on Tariffs and Trade (GATT). Under the Uruguay Round of multilateral trade negotiations, the aspirations of nations seeking an efficient regulatory structure for the application of intellectual property rights was realized, within the establishment of the World Trade Organization (WTO). The Uruguay Round reflected a synergy between States in the establishment of a uniform regime that would harmonize intellectual property rights within member States of WTO. The resulting Trade Related Activities of Intellectual Property (TRIPs) agreement focuses on linking the protection of intellectual property rights to the promotion of innovation in technology and the sharing of that technology in a manner facilitative of social and economic progress. Although it is arguable from a perspective of applied economics that there is an identifiable link between outer space activities and human welfare, nonetheless the basic principle embodied in the TRIPs agreement pertaining to technological innovation would indeed be relevant to activities being carried out in outer space.

[36]WIPO coordinates and administers 22 multilateral unions and conventions on intellectual property protection and sets standards for domestic laws of its members.

TRIPs came to light as a result of an effort by the global community to provide holders of intellectual property rights with an effective mechanism to combat piracy and ensure progressive and equitable trade practices throughout the world.[37] The justification for TRIPs is reported to lie in the existing need to encourage the people of the world to hone their creative and inventive skills toward the betterment of the world, and to this extent the relevance of IRPs to outer space activities cannot be denied. Another important issue for outer space activity in this regard is that the TRIPs agreement, although retaining primacy of objective in the protection of intellectual property rights, is also calculated toward promoting technological innovation and the transfer and dissemination of technology.[38]

In terms of the territoriality of an outer space object or space station, Article 1(1) of TRIPs gives legal legitimacy to a State deciding to ascribe its sovereignty to modules or objects belonging to that State by providing:

> ... Members shall be free to determine the appropriate method of implementing the provisions of this agreement within their own legal system and practice.

This provision not only grants member States a certain discretion in interpreting and applying the TRIPs principles from a local perspective, but it also may provide, as one commentator argues, *sui generis* protection to inventions that may not merit patent protection. This provision also accomplishes, *in limine* the establishment of a link between extra territorial and territorial application of intellectual property rights by a member State by granting the flexibility of extending its local legislation to patents in outer space. This provision also ties in logically with Article VIII of the Outer Space Treaty (already mentioned) and "exports" with some justification TRIPs to outer space activities should a State wish to do so.[39]

Sharing Information and Technology[40]

One of the basic principles enunciated in Space Law and enshrined in the Outer Space Treaty is that space exploration will be for the benefit of all humanity. Article 67 of the TRIPs Agreement has a similar provision which stipulates that:

> In order to facilitate the implementation of this Agreement, developed country members shall provide, on request, and on mutually agreed terms and conditions, technical and financial cooperation in favour of developing and least developed country members.

---

[37]See Final Draft Position Paper on TRIPs, WIPO and WTO, EU Committee of the American Chamber of Commerce in Belgium, 21 May 1999 Brussels.

[38]See Final Draft Position Paper on TRIPs, WIPO and WTO, EU Committee of the American Chamber of Commerce in Belgium, 21 May 1999 Brussels, p. 3.

[39]*35 USC Sec 105* extends US patent laws to outer space.

[40]OECD has defined "technology" as "the systematic knowledge for the manufacture of a product for the application of a process or for the rendering of a service, including any integrally associated managerial and marketing techniques". See OECD (1981), p. 18.

Although the tasks within the common objective differ, in that the intent of the TRIPs Agreement is to impose obligations on developed States to assist other States in the preparation of intellectual property laws and related issues thereto, the objective and principle enunciated in the Outer Space Treaty, of sharing information and technology would be rendered nugatory and destitute of effect if some States were to be "uninitiated" to the process of protection of such information and technology. Furthermore, Article 66(2) of the TRIPs agreement stipulates:

> Developed country members shall provide incentives to enterprises and institutions in their territories for the purpose of promoting and encouraging technology transfer to least-developed country members in order to enable them to create a sound and viable technological base.

This provision gives further thrust to the principles of cooperation in outer space activities which are encouraged between outer space faring States and other States.

One commentator[41] argues that the TRIPs Agreement militates against the economic interests of developing nations since developing nations are precluded from obtaining the "soft" protection earlier afforded to them by WIPO.[42] He argues that technological development reflects the aims and aspirations of developed nations and western needs and standards which developing nations are forced to follow irrespective of the deleterious effects these developments and their demands may have on their economies.[43]

Intellectual Property Rights Regarding Outer Space Activities

For the first time, mention of intellectual property rights pertaining to outer space activity was made and acknowledgment of the validity of such rights was confirmed at the 51st session of the United Nations General Assembly. The Report of the United Nations Committee on the Peaceful Uses of Outer Space (UNCOPUOS) reflecting work of the 35th Session of the Legal Subcommittee of UNCOPUOS, in Annex IV recommended the States be free to determine all aspects of their participation in international cooperation in the exploration and use of outer space on an equitable and mutually acceptable basis. The Report went on further to suggest that contractual terms in such cooperative ventures should be fair and reasonable. One of the ways in which UNCOPUOS identified "fair and reasonable" cooperation is through compliance with the legitimate rights and interests of the parties concerned such as in the field of intellectual property rights.[44] This principle also goes to support the proposition that although proprietary rights in outer space cannot be enforced by States, there could be a valid recognition of territoriality in activities

[41]Islam (1999), p. 190.

[42]Islam (1999), p. 191.

[43]Islam (1999), p. 191.

[44]See http://www.space-generation.org/faq/index.html#u32.

carried out in outer space if such activities were to take place in modules or equipment that have been registered in a particular State.

In the United States, there are numerous statutory provisions[45] pertaining to or at least embodying in certain instances intellectual property laws that may apply to outer space activity: Executive Order 10096 which provides *inter alia* for protection of intellectual property rights of the US Government concerning inventions of government employees stipulates that the Government obtains rights *in toto*, title and interests in any and all inventions of a government employee when such inventions are designed or produced during the scope of employment of the employee concerned.

A memorandum addressed to the Heads of Executive Departments and Agencies by the United States Government on 18 February 1983 lays down governmental patent policy pertaining particularly to rights concerning inventions brought about during the course of government research activity and implementation of development contracts. This memorandum is particularly relevant to the activities of the National Aeronautics and Space Administration (NASA) which, although strictly not binding on the Administration exhorts NASA to comply with the memorandum in accordance with the spirit of the document. It expresses the expectation that NASA (and other similar entities) will make optimum use of the flexibility made available to them to comply with the memorandum.[46]

Unlike the United States, Canada does not have statutory instruments pertaining to inventions made in outer space or resulting from activities in outer space. However, Canada has adopted the Canadian Space Agency Act of 1990 which established the Canadian Space Agency responsible for outer space activities carried out by Canada. Since the Canadian Space Agency is an instrumentality of State having the status of a department of the Federal Government under the Ministry of Industry, employees of the space agency are governed by the Public Servants Invention Act, Section 3 which provides a list of inventions covered by the Act and provides further that all rights pertaining to an invention made by a public servant while acting within the scope of his duties or employment or made by a public servant with facilities, equipment or financial aid provided by or on behalf of Her Majesty the Queen are vested in Her Majesty in right of Canada.

Intellectual property rights are usually acquired on a territorial basis either directly by the inventor or his legal representatives or assignors or by way of transfer, by way of purchase or grant of licence. From an international perspective,

---

[45]See, 42 U.S.C. ξ 2457 pertaining to property rights in inventions; 37 CFR Part 501 on uniform patent policy for rights in inventions of government employees; Executive Order 10096 providing for a Uniform Patent Policy for the US Government; 37 CFR Part 401 on rights to inventions made by non profit organizations and small business firms under government grants, contracts and cooperative agreements; 37 CFR Part 404 on licencing of government-owned incentives; and 35 U.S.C. ξ 105 on inventions in outer space.

[46]It must be noted that NASA, which is an instrumentality of the US Government, has its own guidelines pertaining to the acquisition and control of intellectual property rights. See NASA Act (1958) Sec. 305 as amended by 42 USC ch.26: 1 ξ 2457.

the *Patent Cooperation Treaty of 1970*, which has been ratified by most industrialized nations, makes provision enabling applicants wishing to register their patent in several States at the same time. Section 53(1) of the Act requires that a petition for a patent must be truthful and a false statement, even innocently made, could render such application nugatory and invalid.[47] As long as the application for a patent is made for a right invention for the right owner, misstatements pertaining to various immaterial facts, such as the correct name of the applicant's employer or a different name given to an invention (provided the application is granted regarding the correct invention) are irrelevant and immaterial.

The *Patent Cooperation Treaty* does not require that patents need to be for inventions made for public benefit. However, natural phenomena, such as new species of animals discovered or life forms observed, are not patentable. These are categorized as natural phenomena whether occurring in outer space or on Earth. Schemes, plans, business methods, and even computer programmes in general – the latter being precluded from being registered as a patent in order to preclude the rapid technological progress in the industry – are also not patentable.

In the realm of outer space activity, an intellectual property law that would enforce rights in intellectual property should be primarily calculated to encouraging invention and competition. Although the public interest element in this particular area would be even less than that existing in other areas of intellectual property rights enforcement, States' interest in recognizing the existence of such rights should be essential only if purely for the economic and competitive element involved. Space technology is one of the most sophisticated of technologies and a patent system that would encourage invention, rather than one which acts merely as a rubber stamp, would also be a relevant consideration. The lack of concern of the system for the need for any particular invention is particularly inimical to space science and technology as entities controlling new inventions may, if it were to be to their advantage, not disclose their inventions until they choose to do so. Firms can hide their technology and protect their inventions through principles of law applicable to trade secrets.

The protection of intellectual property rights at space law should contain or identify clear principles of infringement. These should be established on a balance between economic theory and social justice, making sure the protection of intellectual property rights would not only benefit the inventors but also those who later improve, enlarge and challenge inventions already made. For this, an optimum balance is needed between the interests of the inventor, the State concerned, and those who improve space technology.[48]

Of critical importance is the need to introduce some stimulus toward encouraging invention and, at the same time protecting attendant rights.

---

[47]See the Canadian cases of *Beloit Canada Ltée/Ltd.* v. *Valmet Oy* (1984), 78 C.P.R. (2d) 1 at pp. 28–29 (Fed. T.D).

[48]*Hilton-Davis Chemical Co.* v. *Warner-Jenkinson Co. Inc.* 62F 3d. 1512 at 1531–1532 (Fed. Cir. 1995).

The Use of Satellite Imagery at Pre Trial and Trial Hearings

On 3 February 1998, a U.S. Marine EA-6B Prowler Jet conducting a low-altitude training mission near Cavalese, Italy, hit a cable supporting a gondola transportation system being operated across a wide valley. The impact of the jet on the cable severed it, immediately cutting off support for the gondola which plummeted to the ground nearly 600 feet. The accident killed 20 people.

The only manner in which a pre trial presentation could show the jury exactly what occurred during the accident was by recreating the incident. The Italian Government absolutely refused the defence attorneys, who had obtained a jet to recreate a flight path in the same valley, in order to recreate the incident. The only viable alternative available to the defence attorneys was to recreate the flight path with computer-based simulation, which was offered by a firm called Visual Forensics. Although such recreation through digital technology was possible using a helicopter-based global positioning system through a recording of a low speed flight through the valley followed by a digital recalibration of the film simulating the jet's air speed of 600 m.p.h., by the time Visual Forensics were ready for the digital reconstruction, snow and ground conditions had been rendered significantly different from the photographic record of the accident scene.

Another debilitating factor was that no one was able to map out with precision the exact flight path during the critical last phase of the jet's approach. The jet's black box was not sophisticated enough to provide spatial coordinates to pinpoint the aircraft's precise altitude and trajectory.

The entire accident, to the most precise detail, could have been captured in earth photos taken from space platforms which now reveal precise and sharp spatial, temporal and spectral information about happenings on earth. This is accomplished by satellites circling the earth from 400 miles in space at 16,000 miles per hour and having the capacity to transmit data to the ground at high speed rates. Assisted by the global positioning system (GPS), these satellites are able to determine their orbital position in a precise and sharp way which permits them to position with high accuracy ground features for mapping and other applications within a few meters. Additionally, high-resolution space cameras which have the capability of multi-spectral (blue, green, red and near infra-red) imagery, have the capacity and ability to identify objects positioned on the ground which have hitherto been invisible to the human eye. This incredibly sophisticated technology makes detection and identification of objects as small as one meter now possible. The ultimate result could be that satellite imagery could not only be invaluable for agricultural, environmental, land use, hydrocarbon exploration and disaster assessment but it could also be an innovative evidentiary tool for litigators.

The Commercial availability and capability of production of higher resolution spaceborne imagery is now establishing space imagery as a compelling tool in litigation. Therefore, a legal community poised to enter the frontiers of space age technology will find in space imagery cogent and clear evidence analogous to DNA evidence that is now being used in the forensic determination of bio medical identification.

Satellite Imagery

Satellite imagery involves a process which uses cameras or sensor systems usually mounted on an orbiting satellite that capture light reflected from the earth's surface. The fundamental principle of this imagery process is that natural and man-made materials absorb and reflect varying quantities of light in different wavelengths and, through this absorption and reflection process, energy (light) enables satellite sensors to collect data by interacting with objects on the surface of the earth such as plants, soils and buildings in a way that makes possible the extraction of such data and information. Such information is absorbed by the camera/sensor system on the orbiting satellite and transmitted back to earth in a digital format and transformed or converted to images that are capable of being interpreted by sophisticated image processing software. This process makes satellite imagery ideally applicable to the requirement for detail pertaining to large area coverage such as regional cartography, environmental assessment, infrastructure planning and agricultural monitoring.[49]

A significant development in recent satellite imagery through spatial information is the launching of the highest resolution commercial remote sensing satellite with an on-board digital camera having the capability of producing one-meter or better resolution at nadir to focus on the lowest point in the panchromatic, full spectrum mode and four-meter resolution imagery in the red, green, blue and near infra-red (NIR) bands. Such a one-meter resolution will enable the user to distinguish between objects which are one meter in size on the ground if they have distinguishable physical and visual characteristics. Easily detectable under this spatial recognition system are stripes in parking lots, swimming pools, cars, trucks, boats, tennis courts, landscape features, all within their surroundings and environs.

Satellite imagery is not restricted to photography. New sophisticated electronic sensors enable even more powerful imagery that mere photos – digital data that can be analysed, processed and interpreted on computers – computer analysis in digital format admits of possibilities for data to be used in multifarious ways using differences in spectral responses from ground features. The usefulness of this process is emphasized and highlighted when satellite imagery is used in conjunction with the geographic information system (GIS) and the global positioning system (GPS), which together can construct a complete model of an area.

The dynamic applications of a spatial satellite imaging system will be invaluable to attorneys, insurers and risk managers in acquiring data and information on natural and man-made disasters. The Chernobyl Nuclear Plant disaster and the Exxon Valdez oil spill in Alaska are two such incidents which were tracked and recorded for use by television and major newspaper media. The most significant area of contribution is in prediction or preparation where, a satellite just overhead of a disaster area can identify through its imagery an impending disaster or occurrence.

[49]See KPMG. Marwick (1998), p. 5.

Space Law Applications

The essential activity which forms a basis for satellite imagery is remote sensing. Remote sending is the use of satellite technology in obtaining pictures of a territory on earth. This activity is covered by the Outer Space Treaty of 1967[50] which came into force on 10 October 1967. This treaty obliges signatory States to inform the United Nations Secretary General, the public and the international scientific community of the "nature, conduct, locations and results" of space activities.[51] The Space Treaty also imposes international responsibility upon signatory States regarding private sector activities conducted within their territories resulting in activities in space by such private sector entities.[52] This, notwithstanding the collection of satellite images through remote sensing or other technology, is not prohibited under international law since no international treaty contains any specific provision precluding the satellite imaging of objects or places on earth. Moreover, there is no known definition of remote sensing which is essentially the science or art of obtaining information about an object, area or phenomenon through the analysis of data acquired by a device that is not in contact with the object, area or phenomenon under investigation.[53]

At the incipient stage of remote sensing applications, the deployment of remote sensing satellites were for military strategic purposes, remote sensing can now be used to further scientific and discovery objectives. Commercial pursuit of remote sensing for scientific and discovery objectives, called civil remote sensing, is usually conducted by agencies, organizations and individuals who exploit remote sensing systems and data to promote the general welfare and provide for the public good.[54] Commercial remote sensing may be impelled by the motive to make profit, either through procurement or sale of data obtained through sensing.[55]

A legal regime applying a codification of customary legal principles calculated to bind nations and private enterprises through nations was set up in 1987 by the United Nations, when it adopted Resolution 41/65.[56] This resolution was a carefully thought out one, which took the United States and other nations involved 13 years

---

[50]Treaty on Principles Governing the Activities of States in the Exploration and Use of Outer Space, the Moon and other Celestial Bodies, 27 January 1967, 610 U.N.T.S. 205, 18 U.S.T. 2410, T.I.A.S. No.6347, 6.I.L.M. 386 (entered into force on 10 October 1967).

[51]Treaty on Principles Governing the Activities of States in the Exploration and Use of Outer Space, the Moon and other Celestial Bodies, 27 January 1967, 610 U.N.T.S. 205, 18 U.S.T. 2410, T.I.A.S. No.6347, 6.I.L.M. 386 (entered into force on 10 October 1967). Article XI.

[52]Treaty on Principles Governing the Activities of States in the Exploration and Use of Outer Space, the Moon and other Celestial Bodies, 27 January 1967, 610 U.N.T.S. 205, 18 U.S.T. 2410, T.I.A.S. No.6347, 6.I.L.M. 386 (entered into force on 10 October 1967). Article VI.

[53]See Lillesand and Kiefer (1979), p. 1.

[54]Johnson et al. (1993), p. 2.

[55]Johnson et al. (1993) at p. 2 and 3.

[56]United Nations Principles Relating to Remote Sensing of the Earth from Space, G.A. Res 41/65 (XLII), UN GOAR, 29 Sess. 95th Plen. Mtg., UN Doc A/RES/41/65 (1987) ann. at p. 2.

to draft, beginning 1974. The remote sensing principles, as established by Resolution A41/65 admit of access and distribution of data and information generated by remote sensing systems. This activity has its genesis in the Outer Space Treaty of 1967 which stipulates that the exploration and use of outer space, including the moon and other celestial bodies, shall be carried out for the benefit of and in the interest of all countries, on the basis of equality. Article IX of the Treaty embodies a principle of cooperation between nations and provides that the exploration and use of outer space shall be carried out by States parties in a spirit of cooperation and mutual assistance and with due regard to the corresponding interests of all other parties to the Treaty.

There is also explicit recognition that principles of international law apply to space law. The General Assembly of the United Nations in 1961 adopted the view that international law, including the Charter of the United Nations, applies to outer space and celestial bodies.[57] It is also now recognized as a principle of international law that the breach of a duty involves an obligation to make reparation appropriately and adequately. This reparation is regarded as the indispensable complement of a failure to apply a convention and is applied as an inarticulate premise that need not be stated in the breached convention itself.[58] The ICJ affirmed this principle in 1949 in the *Corfu Channel* Case[59] by holding that Albania was responsible under international law to pay compensation to the United Kingdom for not warning that Albania had laid mines in Albanian waters which caused explosions, damaging ships belonging to the United Kingdom. Since the treaty law provisions of liability and the general principles of international law as discussed complement each other in endorsing the liability of States to compensate for damage caused by space objects, there is no contention as to whether in the use of nuclear power sources in outer space, damage caused by the uses of space objects or use thereof would not go uncompensated. The rationale for the award of compensation is explicitly included in Article XII of the *Liability Convention* which requires that the person aggrieved or injured should be restored (by the award of compensation to him) to the condition in which he would have been if the damage had not occurred. Furthermore, under the principles of international law, moral damages based on pain, suffering and humiliation, as well as on other considerations, are considered recoverable.[60]

As discussed, both treaty law and general principles of international law on the subject of space law make the two elements of liability and responsibility a means to an end - that of awarding compensation to an aggrieved State or other subject under the law. Therefore, in view of the many legal issues that may arise, the primary purpose of a regulatory body which sets standards on State involvement in

[57]Resolution 1721 (XVI) adopted on 20 December 1961. See also Article 3 of the *Outer Space Treaty*.

[58]*In Re. Chorzow Factory (Jurisdiction) Case (1927) PCIJ, Ser. A, no. 9* at 21.

[59]*ICJ Reports (1949)*, 4 at 23.

[60]Christol (1991), p. 231.

issues concerning the use of space technology would be to carefully consider the subtleties of responsibility and liability and explore their consequences on States and others involved as they apply to the overall concept of the status of a State as a user of space technology which may cause harm or injury to the latter.

Case Law

The most significant contribution of satellite images as evidence is through their detail and their availability in close proximity to an accident or other litigation causing event. Satellite images have been used in the United States from 1974. In *United States* v. *Reserve Mining*[61] satellite images were admitted into evidence to illustrate the dispersion of taconite tailings by a mining company. In 1988, the Courts, while rejecting the evidence of an expert witness that flooding had increased in a particular area, accepted satellite imagery on the same point as evidence.[62]

There are two ways in which jurisdictions in the United States accept satellite images as evidence: as demonstrative evidence; and as scientific evidence. In the first instance, those images are largely recognized as evidence that would sum up or calculate voluminous evidence which cannot conveniently be examined in Court.[63] This judicial approach accommodates space imaging's CARTERRA product which generates maps, reports and three dimensional images from geographic information systems (GIS) and remote sensing tools. The CARTERRA product reduces complex, voluminous data into scientifically accurate and sharply illustrative charts. The principle of accepting illustrative charts as evidence that could be more illustrative and enlightening to a jury was recognized in 1965 in the case of *McDaniel* v. *U.S.*[64] In the 1984 case of *People* v. *McHugh*[65] graphic computer presentations were accepted by the Court as being a chart or diagram acceptable in evidence. The *McHugh* case was the first instance where a graphic computer presentation was admitted as evidence in a criminal trial and the Court recognized in this instance that computers are simply mechanical tools and when the outcome of a computer analysis is useful and clear it should be accepted, while any computer output which was confusing should be rejected.

Satellite images are also considered in some instances as scientific evidence when they corroborate or prove an opinion given by an expert or when they provide

---

[61]380 F. Supp. 11 (D-Minn. 1974).

[62]*Gasser* v. *United States* 14 Cl.Cf.476 (1988).

[63]Federal Regulation 1006 provides: "the contents or voluminous writings, recordings or photographs which cannot conveniently be examined by Court may be presented in the form of a chart, summary or calculation." See also Hodge (1997), p. 691.

[64]343 F. 2d. 785 (5th Circ. 1965).

[65]124 Misc. 2d. 559, at 560 also 476 N.Y.S. 2d. 721, at 722.

sufficient data for an expert to reasonably base an opinion or conclusion.[66] However, in the 1993 case of *Daubert* v. *Merrell Dow Pharm., Inc.*[67] the trial Court emphasized that such evidence will be admitted with caution, where the Court would act as "gate keeper" and will ensure that evidence so submitted will be not only relevant but also reliable.[68] The Court in the *Daubert* case added that the admissibility of remote sensing information must be examined within the context of the general requirements for admission of scientific evidence and expert opinion. A year later, the Tenth Circuit in *Robinson* v. *Missouri Pacific R.R.Co.*[69] interpreted the *Daubert* principle to be flexible, when it decided to admit a computer animation which demonstrated an expert's opinion relating to the cause of a railroad accident. According to the Appellate Court in *Robinson*, any scientific evidence should have as its overarching objection the scientific validity of evidence presented and must establish the evidentiary reliability and relevance of the principles that underlie a proposed submission.

Opponents of the use of satellite imagery may argue that such imagery is not admissible on the ground and that they constitute hearsay evidence . This can be obviated by conjoining satellite imagery with an expert's testimony since remote sensing data is transmitted to earth in digital format prior to being converted to an interpretable picture, requiring an individual to enter data to produce a recognizable satellite image. In *United States* v. *Elkins*[70] it was held that expert testimony may well include hearsay evidence if the basis of the testimony is relied upon to a reasonable degree by members of the expert's profession.

In order to admit satellite imagery at a trial, litigators have to qualify an expert to introduce and explain such imagery; authenticate and prove the contents of data carried in the images; and establish that proper and accepted digital imagery processing techniques were employed. The last two criteria are stringently relied upon by the Courts which recognize that satellite imagery can be manipulated just as photographs, video tapes and computer simulations which generate visual evidence can.[71] In addition, the expert must also establish to Court that the computer used in processing the satellite images was functioning properly; the scientific analysis used was sufficiently accurate and comprehensive; and the data were relevant and reliable.

---

[66]Federal Evidence Rule 702 provides that: "if scientific, technical or other specialized knowledge will assist the trier of fact to understand the evidence or to determine a fact in issue, a witness qualified as an expert by knowledge, skill, experience, training or education may testify thereto in the form of an opinion or otherwise".

[67]509 U.S. 579 at 590.

[68]See also, *Kumho Tire Co.* v. *Carmichael*, U.S., 119 S.ct.1167 at 1170 (1999) and *Pittston Co.* v. *Allianz Ins. Co.* 805 F.Supp. 1279, at 1370 (D.N.J. 1995).

[69]16 F. 3d. 1083 (10th Circ. 1994).

[70]885 F. 2d. 775 (11th Circ. 1989).

[71]See Hodge (1997) 234 at 717, 727–728.

The satellite imagery submitted as evidence must also be authenticated, i.e., the person bringing forth the evidence must prove that it is what it purports to be.[72] Establishment of this fact is contingent upon the proper handling of data by the satellite data collection company and the transporter of the data and use of an approved scientific method by the expert who interprets the data.[73] Also, a litigator may use similar evidence, such as aerial photography or maps taken on the ground, to authenticate the accuracy and veracity of satellite images. Although distortion of imagery has to be addressed, generally the Courts have accepted that distortion would be remote if the criteria set, as discussed above, are met. The 1999 case of *Dolan* v. *Florida*[74] is a good example, where the Court admitted a computer enhanced image of a surveillance video camera which showed the accused battering a store clerk. Although the tape was of poor quality, the image of the accused and his characteristics were found to be sufficiently clear. It resulted from a process of weighing competing proposals based on different analogies and converging on the view that one of the underlying analogies was better than the other.[75]

Be that as it may, it is not possible to dismiss the pervasive influence exerted by the international legal order in the late fifties when space law was being promulgated. It was this influence which made it possible for the philosophy of space law to be primarily founded upon the concept that, unlike in the formulation of air law in 1944 where States were aware of the economic and technical implications of a tried and tested product, outer space was an unknown quantity, controlled exclusively at that time by two power blocs who had ominous war power, and therefore its philosophy was destined to spread across the globe equally, ensuring commonality and responsibility.

The final consideration pertains to the relevance of satellite imagery as evidence. Purely because of their unique characteristics, satellite images would be deemed to be relevant as evidence having any tendency to make the existence of any fact that is of consequence to the determination of the action more probable or less probable than it would be without the evidence.[76] Nonetheless, a Court will probably carefully examine the evidence to ensure that its probative value is not unduly affected and outweighed by possible unfair prejudice, and confusion of issues. Also important are that there is no undue delay, waste of time, unnecessary presentation of redundant evidence and misleading of the jury. On the last point, Courts have exercised caution in distinguishing between evidence which establishes facts and evidence which merely demonstrate facts. In *Datskow* v. *Teledyne Continental Motors,*[77] the jury was shown a computer animated video to demonstrate how a fire could start and spread in a plane. The Court expressed confidence that the jury

[72]U.S. Fed. R. Evid. 901 (B)(9).

[73]Powell (1996), p. 577 at 585.

[74]1999 WL 512093 (Fla. App. 4th dist.).

[75]Peterson (1997), p. 245 at 266.

[76]U.S. Fed. R. Evid. 40.

[77]826 *F.Supp* 677 (W.D.N.Y. 1993).

would not give the video undue weightage than it deserved. The judge held that the video did not purport to recreate the accident but that it merely demonstrated an expert's view or theory of what may have caused the accident. The conclusion reached by the Court was that, as long as the distinction was made clear to the jury, there was no cause for concern as to misleading the jury.[78] However, a satellite image which would purport to show different ways in which an event may occur may be considered unduly prejudiced. The principle in *Pino* v. *Gauthier*,[79] where a video animation demonstrating four possible ways in which a car could travel over an expressway was rejected as being prejudicial, would be relevant in this instance as precedent.

There is no room for doubt that, in the formulation of space law as a new legal regime, the international community has paid attention to the most significant influence over this discipline of the two superpowers which existed at the time. The two great power blocs, the United States of America and the Union of Soviet Socialist Republics were then the main protagonists (supported by their allies and a few neutral European States and the Third World). This polarization created an environment whereby an ideological balance was required to counter the possibility of encroachment by Realist thought which was calculated to establish an outer space legal regime that would have admitted of the creation of real property rights by one power bloc over the other in outer space to the detriment of humankind in general.

The detriment envisaged was not only in its physical attributes and apprehension for the physical welfare of humanity, but also in the economic imbalance a Realist approach to space law would have brought to bear. The segmentation of space law into municipal regimes would have resulted in the inexorable endorsement of Realist theory, that States are egotistical utility enhancers and that the freedom of internal legislation would have given States a legal licence to maximise their freedom of outer space activities, thereby obviating any possibility of cooperation with other States. The inevitable corollary to this scenario would have been the avoidance of spreading the benefits of outer space exploration and exploitation among the world.

The general delineation of outer space as a common area rather than as an area which could be conquered therefore rested on more traditional ground of international cooperation. However, the reason for the two power blocs to settle on commonality rather than on appropriation as the basis for the philosophy of space law is by no means the result of hard bargaining. It was the preferred choice of both parties. History records that there was no coercion or imposition of one system over another at the time of law making. One commentator says:

> Satellite image production commences when digital data are transmitted to the receiving station on the ground from a satellite and are recorded on a magnetic tape. Data so received

---

[78] 826 *F.Supp* 677 (W.D.N.Y. 1993) at p. 685, see also In *re Air Crash Disaster* 86 F. 3d. 498 (6th Cir. 1996).

[79] 633. So. 2d. 638 at p. 652 (La. App. 1st Cir. 1993).

is adjusted for atmospheric interference and corrected for geometric precision. These processes are necessary and are deemed to be appropriately carried out and if satellite images were to be accepted as evidence.

The proliferation of satellite imagery and its prodigious use as an effective source of spatial, temporal and spectral information makes images taken from outer space burgeoning tools for litigators in their quest for success in the trial process. The most fundamental driver in this process is the acceptance by Courts of satellite imagery as a credible source of information. As explained in this article, there are a few impediments that may preclude this particular kind of evidence as being admitted within the purview of traditional rules of evidence. Questions of authentication, reliability and relevance would be critical issues in this process, along with transparency. However, given the accuracy and safeguards now inherent in sophisticated technology and the advanced expertise of the professionals who analyse and interpret satellite images, this type of evidence may well be widely accepted in the future.

The freedom of the use of outer space would be meaningless if one were to reject its many benefits. If the space odyssey has already arrived in the Courts of law, it should not be an unwelcome visitor.

References

Abeyratne RIR (1992) The philosophy of air law. Am J Jurisprudence 37:135–144

Böckstiegel K-H (1983) Prospects of future development in the law of outer space. Ann Air Space Law VIII:305

Brownlie I (1990) Principles of public international law, 4th edn. Clarendon Press, Oxford, p 433

Cheng B (1968) The 1967 space treaty. J Droit Int 95:578

Christol CQ (1983) The Jus Cogens principle and international space law. XXVI Colloq 26:1

Christol CQ (1991) Space law past, present and future. Kluwer Law and Taxation Publishers, Deventer, p 231

Goedhuis D (1976) Some substantive and procedural issues presently at stake in space legislation. Z Luft Weltraumrecht Ger J Air Space Law 25:198–199

Gorove S (1983) Liability in space law: an overview. Ann Air Space Law VIII:373

Hodge SH (1997) Satellite data and environmental law; technology ripe for litigation application. Pace Env Law Rev 14:718

Islam MR (1999) International trade law. Law Book Company, Sydney, p 190

Jakhu RS. In: Girardot et al (ed) Developing countries and the fundamental principles of international space law. p 351

Johnson DJ, Nelson M, Lampert RJ (1993) U.S. space-based remote sensing: challenges and prospects. Rand, Santa Monica, p 2

Lachs M (1972) The law of outer space: an experience in contemporary law making. Sijthoff, Leiden, p 70

Lauterpacht H (1955) International law, Kluwer, 8th ed. p. 11

Lillesand TM, Kiefer RW (1979) Remote sensing and image interpretation. Wiley, New York, p 1

Markoff MG (1976) Disarmament and 'peaceful purposes' provisions in the 1967 outer space treaty. J Space Law 4:3

Marwick P (1998) The satellite remote sensing industry – a global review. KPMG, Washington, p 5

Matte NM (1980) Aerospace law: telecommunications satellites. Recueil Cours 166:147
OECD (1981) North-south technology transfer; the adjustments ahead. OECD, Paris, p 18
Peterson MJ (1997) The use of analogies in developing outer space law. Int Organ 51:266
Powell CE (1996) Computer generated visual evidence: does Daubert make a difference? Ga State Univ Law Rev 12:585
Vandoran P (1999) The implementation of the TRIPs agreement. J World Intellect Property 2(1):25

Chapter 9
Conclusion

Amidst all its glamour and glitter, space transportation brings to bear two major considerations. The first is that the development of this mode of transportation should essentially be subject to good governance. The second is that any development of space transportation should not endanger and encroach upon the rights of sovereign States and their citizens. These two concepts are inextricably linked to the overarching message of peace that resonates the outer Space treaties and the wisdom that is needed on a global basis for nations to secure a culture of peace through the benefits of space exploration.

Harold Bloom, Sterling Professor of Humanities at Yale University and a former professor at Harvard, states in his book, "Where shall Wisdom be Found," that there are three criteria that impel him to go on reading and teaching: aesthetic splendor, intellectual power and wisdom. Of these, the last is perhaps the most useful for survival. Wisdom is the ability to make correct judgments and decisions, and remains an intangible quality gained through experience. Often, society tends to attribute wisdom to an action or decision that is determined in a pragmatic sense by its popularity. Some criteria in judging wisdom are traditionalism and how long it has been around, and its ability to predict against future events. Wisdom connotes an enlightened perspective.

A standard philosophical approach to wisdom is that it involves making the best use of available knowledge. However, as with all decisions, a wise decision may be made with incomplete information. The technical philosophical term for the opposite of wisdom is folly. In his *Metaphysics*, Aristotle defines wisdom as knowledge of causes: *why* things are the way they are. Traditional western culture associates wisdom with virtue. For example, in the Roman Catholic church, wisdom (or prudence) stands with justice, fortitude and moderation as one of the four cardinal virtues. The source of these virtues was Plato (427 BC–347 BC), who was an immensely influential classical Greek philosopher. Plato was a student of Socrates and the teacher of Aristotle. He was a prolific writer and was the founder of the Academy in Athens.

There are various interpretations of the word "wisdom" Some religions hold that wisdom may be given as a gift from God. Holists believe that wise people sense, work with and align themselves and others to life. According to this view, wise people help others appreciate the fundamental interconnectedness of life. The

R. Abeyratne, *Space Security Law*,
DOI 10.1007/978-3-642-16702-7_9, © Springer-Verlag Berlin Heidelberg 2011

Jewish book of Proverbs in the Old Testament states "Fear of God is the beginning of Wisdom." Confucius, (551 BC–479 BC) was a famous thinker and social philosopher of China, whose teachings have deeply influenced East Asia for centuries. He stated that wisdom can be learned by three methods: Reflection (the noblest), imitation (the easiest) and experience (the bitterest).

In the context of the above approaches to wisdom, one could draw a distinct link with the application of wisdom to a peace process between two parties. At its most simplistic level, the word "peace" entails the absence of war. The ancient Romans defined peace, *Pax*, as *Absentia Belli*, or the *absence of war.*). Peace is many things to many different people around the world. Invariably, the meaning of the word "peace" changes with context. Peace may refer specifically to an agreement concluded to end a war, or to a lack of external warfare, or to a period when a country's armies are not fighting enemies. It can also refer more generally to quietude, such as that common at night or in remote areas, allowing for sleep or meditation. Peace can be an emotion or internal state. And finally, peace can be any combination of these definitions.

A person's conception of "peace" is often the product of culture and upbringing, as is a nation's concept of peace is essentially influenced by the cultural ethos of its people. People of different cultures sometimes disagree about the meaning of the word, and so do people within any given culture. Peace is not a symbol, peace is a mindset.

The maintenance of longstanding peace between nations ranks among the few great successes of the United Nations. Peace can be voluntary, where potential agitators choose to abstain from disturbance, or it can be enforced, by suppressing those who might otherwise cause such disturbance. One of the most significant and wise initiatives of the United Nations was in 1997, when the UN adopted General Assembly Resolution A/52/13 which alludes to a culture of peace that should permeate values, attitudes and conduct and impel the parties concerned to reject violence and attempt to solve problems through consultation, negotiation and consensus. The UN has gone further, in adopting Assembly Resolution 53/243, which is aimed at ensuring for the children of the world a peace culture through an "International Decade for a Culture of Peace and Non Violence."

A culture of peace would entail different features for different nations, depending on their own cultural history and attitudes of racial, religious and gender tolerance. However, there are some basics that would incontrovertibly apply to any given situation or instance. Firstly, the issue would be to what extent are people educated and sensitized to perceive themselves as part of a peaceful society who would follow rules of conduct leading to dialogue and negotiation instead of taking to the use of force. Peace education should bring to bear the compelling need for a tolerance and non-violence based culture. Some issues that would bear on a culture of peace would be: To what extent are women given equality and a say in governance as well as in day to day living?; To what extent are children and their nurturance valued?; what roles do understanding, tolerance, solidarity and mutual obligation play in a given society? how far does a State adhere to democratic participation and how much credence and reliance is placed on open communication, transparency and accountability?.

Perhaps the most significant indicator of a cohesive peace culture is the extent to which a government encourages and facilitates the practice of human rights. A corollary to this is the government's initiatives and attempts to include all diverse components and groups of a society in its agenda.

The international obligations of a government and its society also play a significant part in the culture of peace in a country. How far are UN resolutions on peace initiatives adopted and adhered to? How much does a State participate in joint programs to eradicate global terrorism? For example, to what extent would a State participate in the International Convention for the Suppression of the Financing of Terrorism, adopted by the United Nations General Assembly, on 9 December 1999, aimed at enhancing international co-operation among States in devising and adopting effective measures for the prevention of the financing of terrorism, as well as for its suppression through the prosecution and punishment of its perpetrators?

The Convention, in its Article 2 recognizes that any person who by any means directly or indirectly, unlawfully or wilfully, provides or collects funds with the intention that they should be used or in the knowledge that they are to be used, in full or in part, in order to carry out any act which constitutes an offence under certain named treaties, commits an offence. The treaties listed are those that are already adopted and in force and which address acts of unlawful interference with such activities as deal with air transport and maritime transport. Also cited is the International Convention for the Suppression of Terrorist Bombings, adopted by the General Assembly of the United Nations on 15 December 1997.

The Convention for the Suppression of the Financing of Terrorism also provides that, over and above the acts mentioned, providing or collecting funds toward any other act intended to cause death or serious bodily injury to a civilian, is an offence. Furthermore, the Convention deems it an offence to cause death or injury to any other person not taking an active part in the hostilities in the situation of armed conflict, when the purpose of such act, by its nature or context, is to intimidate a population, or to compel a government or an international organization to do or to abstain from doing any act.

As defined by the United Nations, the culture of peace is a set of values, attitudes, modes of behaviour and ways of life that reject violence and prevent conflicts by tackling their root causes to solve problems through dialogue and negotiation among individuals, groups and nations. Firstly, a culture of peace requires revising the educational curricula to promote qualitative values, attitudes and behaviours of a culture of peace, including peaceful conflict-resolution, dialogue, consensus-building and active non-violence. Such an educational approach should be geared also to promote sustainable economic and social development by reducing economic and social inequalities, by eradicating poverty and by assuring sustainable food security, social justice, durable solutions to debt problems, empowerment of women, special measures for groups with special needs, and environmental sustainability.

In the context of human rights, the United Nations philosophy is that human rights and a culture of peace are complementary and that, whenever war and violence dominate, there is no possibility to ensure human rights. At the same time, without human rights in all their dimensions, there can be no culture of peace. with regard to equality of the sexes, UN policy dictates that through full participation of women in economic, social and political decision-making, elimination of all forms of discrimination and violence against women, and support and assistance to women in need, there could be an environment where a culture of peace would flourish. Also, democratic participation is an indispensable foundation for the achievement and maintenance of peace and security. For this to be achieved, there should be practices and participation in all sectors of society, a transparent and accountable governance and administration, and effective combat against terrorism, organized crime, corruption, illicit drugs and money laundering. The UN also believes that, in order to abolish war we need to abolish violent conflicts. There is also the need for action geared to transcend and overcome enemy images with understanding, tolerance and solidarity among all peoples and cultures, particularly by learning from societal and cultural differences, through dialogue and the exchange of information, which could be an enriching process.

Another important facet of a culture of peace is freedom of information and communication and the sharing of information and knowledge. Furthermore, measures need to be taken to address the issue of violence in the media, including new information and communication technologies. The UN concludes that the gains in human security and disarmament in recent years, including nuclear weapons treaties and the treaty banning land mines, should encourage us to increase our efforts in negotiation of peaceful settlements, elimination of production and traffic of arms and weapons, humanitarian solutions in conflict situations, post-conflict initiative.

The United Nations has used he abovementioned objective indicators as well as their implications and applied these to several nations in the world. The result, as seen from a survey of over seventy nations, showed that when the indicators were correlated in order to search for underlying factors, four factors emerged as truly inter relative to all indicators. The first factor is democratic development, measured in terms of indicators for press freedom, gross domestic product (GDP), life expectancy, literacy, democracy, human rights and the percentage of women in the legislature. The second factor is equality, measured by the Gini index on income inequality, low homicide rate and to a certain extent human rights. The third indicator is non violent means, which essentially is the inverse of military spending and security threats against the nation and its people, which are linked to the people who are imprisoned. The final factor is nurturance, which is measured by the quality of education provided and related expenditure, tolerance for refugees and to some extent the percentage of women in professional jobs including the legislature.

Joseph de Revera, Professor of Psychology at Clark University in the United States, in a recent article published in the *UN Chronicle*, is of the view that this set of objective indicators could be used to measure the degree to which nations posses a culture of peace. These measures are benchmarks which need to be approached on the basis of a study of strengths and weaknesses of a nation headed towards the goals of creating and nurturing a society that would attain a complete culture of peace.

The most essential indicator, which strikes a good balance between negotiating between powerful stakeholders on the one hand and looking after the people of a nation, is nurturance. The essence of wisdom in governance, as was discussed earlier, is compassion and provision. Provision for the needs of a people striving for quality education, healthcare and recreation. A complete culture of peace could only prevail if those responsible on both sides establish that the fundamental goal in a peace process is to make the people content and raise their quality of living while providing them with opportunities that they rightfully seek.

Governance and the United Nations

Current political and diplomatic problems mostly emerge as a result of the inability of the world to veer from its self serving concentration on individual perspectives to collective societal focus. This distorted approach gives rise to undue emphasis being placed on rights rather than duties; on short-term benefits rather than long-term progress and advantage and on purely mercantile perspectives and values rather than higher human values.

At the heart of international politics is the United Nations. Often one hears statements like "the United Nations failed" (for instance to stop the genocide in Rwanda or ethnic cleansing in the Balkans) or "the United Nations succeeded" (to stop the Iraqi occupation of Kuwait). This misconception obfuscates the complex reality that the United Nations is basically an intergovernmental organization in which the key decisions are made by governments representing States. In other words, the United Nations is empowered by the member States to carry on its tasks and not the other way around. Although the Charter of the United Nations initially provides language starting with "we the peoples of the world" in effect the key players who call the shots in the United Nations system are the governments themselves. For instance, when it is said that the Security Council took a decision, what is meant is that representatives of fifteen States made that decision. This is true also of the various specialized agencies such as the World Meteorological Organization and the International Civil Aviation Organization, both of which have Councils that take the decisions and are representative of member States. Over its sixty years of service to the international community, the United Nations has, through its General Assembly and Security Council adopted numerous resolutions. This article will examine the extent to which the Member States of The United Nations are bound by such pronouncements through a legal analysis of how far the United Nations Organization is empowered by States to adopt such resolutions and directives.

The first step toward such an examination would be inquire into the nature of the United Nations. It is represented and directed by its member States. Therefore, it is incontrovertible that universal participation in the United Nations is indispensable if The UN were to effectively implement the provisions of the Charter of the United

Nations. Sixty years of symbiotic existence have shown that States need the UN no less than The UN needs their membership.

An organization such as the UN is tasked primarily to provide a certain predictability about its members by promulgating norms for the conduct of its Contracting States. Of course not all those norms are binding and not all of them are adopted with the same degree of formality. However, certainly all of them provide guidance to States. This situation has to mesh with the basic inquiry as to whether the UN, as an international organization, has been given direct authority over individuals or States.

The question arises as to whether a contracting State is formally bound by a Resolution of the United Nations, particularly when such a State has no convincing argument that it is impracticable to implement such a resolution. International organizations can generally only work on the basis of legal powers that are attributed to them. Presumably, these powers emanate from the sovereign States that form the membership of such organizations. Therefore, the logical conclusion is that if international organizations were to act beyond the powers accorded to them, they would be presumed to act *ultra vires* A seminal judicial decision relating to the powers of international organizations was handed down by the Permanent Court of International Justice in 1922 in a case relating to the issue as to whether the International Labour Organization (set up to regulate international labour relations) was competent to regulate labour relations in the agricultural sector. The court proceeded on the basis that the competence of an international organization with regard to a particular function lay in the treaty provisions applicable to the functions of that organization and that the determination of such competence would be based on interpretation. In this instance the Court was of the view that, in its interpretation of the ILO treaty, the organization had the power to extend its scope of functions to the agricultural sector. However, this principle of implied extension should be carefully applied, along the fundamental principle enunciated by Judge Green Hackworth in the 1949 *Reparation for Injuries Case* – that powers not expressed cannot freely be implied and that implied powers flow from a grant of express powers, and are limited to those that are necessary to the exercise of powers expressly granted.

The universal solidarity of the United Nations member States that was recognized from the outset at the establishment of the Organization brings to bear the need for States to be united in recognizing the effect of UN policy and decisions. This principle was given legal legitimacy in the *ERTA* decision handed down by the Court of Justice of the European Community in 1971. The court held that the competence of the European Community to conclude an agreement on road transport could not be impugned since the member States had recognized Community solidarity and that the Treaty of Rome which governed the Community admitted of a common policy on road transport which the Community regulated.

It should be noted that the United Nations does not only derive implied authority from its Contracting States based on universality but it also has attribution from States to exercise certain powers. The doctrine of attribution of powers comes directly from the will of the founders, and in the United Nation's case, powers were attributed to the United Nations when it was established as an international organization to administer the provisions of the UN Charter. In addition, the United Nations could also lay claims to what are now called "inherent powers" which give the Organization power to perform all acts that the Organization needs to perform to attain its aims not due to any specific source of organizational power but simply because the United Nations inheres in organizationhood. Therefore, as long as acts are not prohibited in The UN Charter, they must be considered legally valid.

Over the past two decades the inherent powers doctrine has been attributed to the United Nations Organization and its specialized agencies on the basis that such organizations could be stultified if they were to be bogged down in a quagmire of interpretation and judicial determination in the exercise of their duties. The advantages of the inherent powers doctrine is twofold. Firstly, inherent powers are functional and help the organization concerned to reach its aims without being tied by legal niceties. Secondly, it relieves the organization of legal controls that might otherwise effectively preclude that organization from achieving its aims and objectives. The ability to exercise its inherent powers has enabled the United Nations to address issues that are not directly within its mandate but directly or indirectly impact its core functions.

With regard to the conferral of powers by States to the United Nations, States have followed the classic approach of doing so through an international treaty. However, neither is there explicit mention of such a conferral on the United Nations in the Charter of the UN, nor is there any description of the United Nations' powers. Of course the Security Council can impose sanctions on States that are deemed to act inconsistently with the principles of the Charter. Therefore States have not followed the usual style of conferral of powers in the case of the United Nations, which, along the lines of the decision of the International Court of Justice in the 1996 *WHO Advisory Opinion* case was that the powers conferred on international organizations are normally the subject of express statement in their constituent instruments. This notwithstanding, it cannot be disputed that the United Nations member States have conferred certain powers on the UN to perform its functions independently. For example, the United Nations is a legal entity having the power to enter into legal agreements with legal entities including other international organizations with regard to the performance of its functions.

Conversely, an international organization must accept conferred powers on the basis of Article 34 of the Vienna Convention on the Law of Treaties which stipulates that a treaty does not create rights or obligations of a third State without its consent. This principle can be applied *mutatis mutandis* to an international organization such as the UN. The conferral of powers on an international organization does not *ipso facto* curtail the powers of a State to act outside the purview of that organization unless a State has willingly limited its powers in that respect. This principle was recognized in the *Lotus* Case where the Provisional Court of International Justice held that a State can exercise powers on a unilateral basis even while the conferral to the Organization remains in force. The Court held that restrictions upon the independence of States cannot be presumed.

The United Nations' conferred powers enable the Organization to adopt binding regulations by majority decision. However, States could opt out of these policies or make reservations thereto, usually before such policy enters into force. This is because States have delegated power to The UN to make decisions on the basis that they accept such decisions on the international plane. In such cases States could contract out and enter into binding agreements outside the purview of the United Nations even on subjects on which the UN has adopted policy.

Given the United Nations' profile as a self standing legal entity, the Organization would be responsible for its internationally wrongful acts. As to the issue whether a State which has delegated powers to the United Nations would be responsible for the wrongful acts of the Organization, a State is not bound by the Organization's exercise of delegated powers and therefore it cannot be necessarily assumed that such acts would be attributable to the States unless such acts were the effect of the State's own acts or omissions. Article 1 of the *Articles of Responsibility* of the International Law Commission (ILC) expressly stipulates that every internationally wrongful act entails the international responsibility of a State. The State cannot escape responsibility by seeking refuge behind the non-binding decision of an Organization in the case of delegation of powers. This is also the case where a State aids and abets an Organization to perform an internationally wrongful act.

The General Assembly and the Security Council are composed of Contracting States. The General Assembly has delegated activities concerning matters of international security to the Security Council as well as delegating other areas of work to other Councils of the UN such as the Economic and Social Council (ECOSOC) and the Trusteeship Council. Therefore it would not be incorrect to assume that any resolution adopted or decision taken by the UN Security Council can be imputed to the member States of the UN which have delegated powers on these Councils. However, States retain the powers to act unilaterally and they are not bound to comply with obligations flowing from the Organization's exercise of conferred powers. States which have delegated powers on the United Nations have the legal right under public international law to take measures against a particular exercise by the UN of conferred powers which is considered to be *detournement de pouvoir*, *ultra vires* or an internationally wrongful act with which the objecting States do not wish to be associated. A State could also distance itself from the State practice of other Contracting States within the UN if such activity is calculated to form customary international law that could in turn bind the objecting State if it does not persist in its objections.

As discussed earlier, a significant issue in the determination of the United Nations' effectiveness as an international organization is the overriding principle of universality and global participation of all its member States in the implementation of UN policy. This principle, which has its genesis in the Charter of the United Nations, has flowed on gaining express recognition by legal scholars. This is what makes the United Nations unique and establishes without any doubt that the UN is not just a tool of cooperation among States.

In the years to come, citizens of the world will scrutinize both their governments and those of foreign nations whose responsibility it is to ensure good governance and the continuity of the world communications systems. The politician, diplomat and lawyer will increasingly turn toward principles of international law to determine the best course of action in crisis.

The Role of the Secretary General

The role of the Secretary General of the United Nations is driven by Articles 7 and 97 through 101 of the United Nations Charter. Article 7 recognizes as the principle organs of the United Nations: the General Assembly; Security Council; Economic and Social Council; Trusteeship Council; International Court of Justice; and the Secretariat. Article 97 provides further legal legitimacy to the office of the Secretary General by stating that the Secretariat comprises a Secretary-General and such staff as the Organization may require. It goes on to say that the Secretary-General will be appointed by the General Assembly upon the recommendation of the Security Council and that he is the chief administrative officer of the Organization. The Article that follows states that The Secretary-General will act in his capacity of chief administrative officer of the Organization in all meetings of the General Assembly, of the Security Council, of the Economic and Social Council, and of the Trusteeship Council, and shall perform such other functions as are entrusted to him/her by these organs. The Secretary-General is required to make an annual report to the General Assembly on the work of the Organization.

Article 99 of the Charter provides that the Secretary-General may bring to the attention of the Security Council any matter which in his opinion may threaten the maintenance of international peace and security. Here it must be noted that the mandate given to the Secretary General is limited to international peace and security. Article 100 infuses independence and objectivity to the role of the Secretary General by providing that, in the performance of their duties the Secretary-General and the staff of the Secretariat (which have since been extended to the specialized agencies of the United Nations as well) will not seek or receive instructions from any government or from any other authority external to the Organization. They are required to refrain from any action which might reflect on their position as international officials responsible only to the Organization. This is one reason for the Secretary General and United Nations staff to be known as "rootless cosmopolitans." Article 100 goes on to say that each Member of the United Nations undertakes to respect the exclusively international character of the responsibilities of the Secretary-General and the staff and not to seek to influence them in the discharge of their responsibilities. Article 101 states that staff, who are appointed by the Secretary General, are so appointed under the paramount consideration that in the determination of the conditions of service would be the necessity of securing the highest standards of efficiency, competence, and integrity and that due regard will be paid to the importance of recruiting the staff on as wide a geographical basis as possible.

By virtue of Article 7 of the Charter, the Secretary General is expected, at least impliedly, to uphold the aims and purposes of the Charter. One commentator has said that the appointment of a Secretary-General lies on the support given by the five permanent members of the Security Council, namely the United States, United Kingdom, France, The People's Republic of China and Russia who have the power to veto any proposal. Consequent upon the appointment being confirmed, the Secretary-General does not command economic, physical or military power. Instead, he depends on the moral status emanating from the principles of the UN Charter on his credibility and perceived or assumed impartiality to function effectively. The moral and political status of the office of the Secretary General is of paramount importance in determining the role of the Secretary General and his authority would rest to a large extent on the Charter and the fact that these ideals are formally accepted by all members of the world's universal intergovernmental organization whose servant and spokesman he is. The Secretary General is expected to resonate these ideals in public statements and be an embodiment of the UN in his diplomatic activities.

It has been said that the success of the mediation process by the Secretary-General depends to a large extent on his ambitions, strategies, background, training, work. Over the years, Secretaries General of the United Nations have adopted their own style of administering the Charter. Experience, intellect style, and the ability to network both formally and informally among the most influential members of the Security Council and the General Assembly have proved to be virtues of the leadership roles played by some Secretaries General whereas others have not demonstrated these qualities. On the negative side of the Report is Trygvie Lie of Norway who served as the first Secretary-General from 1946 to 1953. He is reported to have antagonized both the United States and the Union of Soviet Socialist Republics and been forced to resign from office. Dag Hammarskjold of Sweden, who was Secretary General from 1953–1961, is reported to have enabled through his inspired leadership the office of the Secretary General and the Secretariat, and to some extent the United Nations itself to build a new profile and exalted status in the international community of nations.

A commentator has said: "He provided the most dynamic and striking leadership the international organization ever had, and he personified the ideals of the Charter in action in a way that made a profound impression on hundreds of millions of people all over the world." His successor, U Thant of Burma who was Secretary General from 1962 to 1971 has been accused of being devoid of the high standards expected of the office. However, others have hailed him as the quiet diplomat, who through his unobtrusive efficiency and effectiveness struck deals in times of conflict which ameliorated many a difficult diplomatic situation. In the case of Kurt Waldheim of Austria, Secretary General from 1972 to 1982, a commentator (Urquhart) has noted that he was "an energetic mediocrity who lacked the qualities of vision, integrity, inspiration and leadership that the United Nations so desperately needs." Javier Perez de Cuellar (1982–1992) who succeeded him has been called a self-effacing diplomat who confounded critics by effectively mediating some of the bitterest conflicts of the decade, thus rescuing the UN precisely in the core area where it had seemed to lose all relevance. He goes down in history as a leader who salvaged the credibility of the United Nations and ensured that the

Organization could play a leading role in the future. He is also widely remembered as one of the most credible and effective leaders of the UN in its first half-century of its existence. The former Secretary General Kofi Annan, who had the distinction of being from the Secretariat for many years prior to his appointment, has been credited with favourable comments as a consensus builder and a polished diplomat par excellence. His rating was very high especially after getting Iraq to sign an undertaking that the country will abide by the resolutions of the Security Council.

Under the Charter, the Secretary General has neither the authority nor the powers to influence the Security Council or a member State. He can give a legal opinion to the Council. An important provision of the United Nations lies in Article 2(7) of the Charter which provides that nothing contained in the Charter would authorize the United Nations to intervene in matters which are essentially within the domestic jurisdiction of any state. It further states that Members of the United Nations are not required to submit such matters to settlement under the Charter; but that this principle would at the same time not prejudice the application of enforcement measures under Chapter Vll of the Charter which addresses the issue of action to be taken by the Security Council to threats to the peace, breaches of the peace and acts of aggression.

Therefore any internal issue within a member nation of the United Nations is, *ipso facto* beyond the jurisdiction of the United Nations. The only exception to this principle is where, in exceptional circumstances the Security Council can interpret action taken by the Secretary General as promoting international peace, which in turn would circumvent Article 2(7). One such example was when U Thant sent observers to the border in the 1963 Yemeni civil war to prevent border infiltration The Security Council found that the matter was not purely internal and therefore within UN's competence.

In the ultimate analysis therefore, the Secretary General has to act strictly within the moral precincts of the United Nations Charter which promotes international peace. Perez de Cuellar once observed that, the aim of traditional diplomacy was often limited to a stable balance of power and, whether the balance conformed to justice was of a lesser concern. He went on to say that peace as envisaged by the UN Charter is a just peace and the Secretary General should have no part in any diplomatic deal or undertaking which ignores the principles of the Charter or relevant pronouncement of the competent organs of the UN.

Space Transportation and Human Rights

Corporations which involve themselves in the business of space transportation – which by itself could be is a multi trillion dollar industry should be cognizant and conscious of their duty to society. One specific area which stands out in the field of human social responsibility is corporate responsibility. The Secretary General of the United Nations appointed in 2006 a Special Representative on this subject, who completed a report on corporate duty towards the citizen and the duty of the State in protecting the citizen from corporate infringement of human rights. This report has compelling and wide ranging implications for global business, in particular for corporate entities that operate business globally. Similar implications can be seen with regard to emerging markets, underdeveloped countries, or countries that still remain destitute of a democratic system. The Special Representative's Report proposed that a certain responsibility to respect human rights devolve upon corporations and that the State has a "duty to protect" its citizens against human rights abuses by companies, and that both the State and businesses must provide more effective access to remedies for human rights violations.

Although we now have the assurance that corporations' duty to respect human rights essentially means not to infringe on the rights of others or, in other words, to do no harm, the abovementioned philosophy which was recommended to the United Nations may well place on businesses numerous expansive obligations that require close attention by corporate management and boards. Under this philosophy, corporate entities would be required to conduct a broad due diligence process to become aware of, prevent and address adverse human rights impacts, purportedly in the same way as corporations already must assess and manage financial and related risks in order that they discharge their responsibility to respect human rights. Essentially, the philosophy contained in the Report of the Special Representative to the Secretary General of the United Nations would impose on corporations the obligation to compensate for the political, civil, economic, social, or other deficiencies of the countries in which they conduct business. Further, corporate boards of directors may even be expected to monitor and ensure the vindication of broad-textured principles enshrined in various international human rights instruments.

Compensation for Damage Caused by Space Transportation

The proliferation in the times to come, of space transportation would bring to bear issues of responsibility and liability of those responsible. The award of compensation would be the natural recourse under principles of tort law in the common law context.

It becomes clear that the enormity of such disasters, both in extent and consequences, brings to bear the need to view the principles of negligence in a new light. In most globalized economies, tort law,[1] which is the branch of law that provides compensation for injuries to persons and property caused by the act of another, is a constantly evolving area of the law. This continuous evolution is caused by new and emerging social and economic activities brought about by technological advancement and increasing and varied commercial activity. One

[1]See White (1980), XI.

of the inherent difficulties in dealing with tort law is that it has defied definition.[2] However it has two determinants: actual or legal damages caused to the plaintiff by the act or acts of the defendant[3] and the fact that the act of the defendant could be determined on the basis of fault liability[4] or strict liability.[5]

The private sector in any society drives the economy by bringing new opportunities for employment, technology and capital, and paving the way for development and improvement of work conditions and standards of living. One commentator has put forward the view that courts must make directors and senior officers of companies liable for intentional torts that affect others in the course of their duties.[6] A natural corollary to this approach would be to consider the employer as the insurer of the employee whereby the latter would be compensated for a tort by holding the directors and officers of a company jointly and severally liable.[7]

Private companies are powerful anywhere. Gabel and Bruner, in their book *Global Inc,: An Atlas of Multinational Corporations*,[8] say that the three hundred largest corporations account for more than one quarter of the worlds productive assets. Ninety million people are employed by transnational corporations (twenty million of whom live in developing countries). These companies produce twenty five percent of the world's gross product and the top one thousand of these companies account for eighty percent of the world's industrial output.

Voluntary codes of conduct have been adopted by several industry associations. However, these codes tend to be highly conceptual and do not lend themselves to easy application. A recent study conducted in this area suggests that only eighty five corporations have even mentioned human rights in their company codes. A good starting point for any business entity that is interested in ensuring human rights in the workplace are the *Norms on the Responsibilities of Trans National Corporations and Other Business Enterprises with Regard to Human Rights* approved by the UN Sub Commission on the Protection of Human Rights in August 2003. The Norms have five significant attributes that are relevant, the first being that nothing in the Norms would diminish the primary role of the State in ensuring human rights of its citizens and the pre-eminent obligation of the government in that regard. Perhaps the most important contribution of the Norms is to clearly establish that they apply not only to transnational corporations, but also to national companies and local businesses. By this measure, the Norms ensure that they are applied

---

[2]Prosser (1964) at 1.

[3]Dudley and Baylies (1876), at 11.

[4]Fault liability is based on the defendant's conduct, where the plaintiff proves that the defendant was guilty of wrongful conduct which is either intentional or negligent. See Osborne (2003) at 24.

[5]The distinguishing feature of strict liability is that the plaintiff does not have to prove the guilt of the defendant of any wrongful or negligent conduct. Mere proof that the defendant caused the plaintiff's loss in the manner prescribed is sufficient to impose liability. See *Rylands* v. *Fletcher*, (1868), L.R. 3 H.L. 330.

[6]See Feasby, Corporate Agents Liability in Tort, 199 32 *C.B.L.J.* 291 at 298.

[7]See *Proctor* v. *Seagram*, [1925] 2. D.L.R. 1112 at 1114.

[8]The New Press: New York 2003 at 34.

without discrimination, whilst obviating the possibility of business entities skillfully seeking exemptions brought to bear by the type of organization as defined or made open to interpretation. Furthermore, this approach creates a harmonious balance between all businesses however large or small, while not being too onerous on small businesses.

The Norms are far reaching and generally encompass spectrum of human rights spread out over twenty three paragraphs. They include the right to equality of opportunity and treatment; the right to security of persons; the rights of workers; the right to collective bargaining; respect for the rule of law and international and local laws; the right to a healthy work environment; the right to political and social and cultural rights; and other civil rights. The Norms are by no means laws in the nature of treaties or other international legal instruments, but remain as guidelines in the nature of other UN declarations, principles and standards. The most effective feature of the Norms is its implementation process which details five basic implementation procedures. Firstly, the Norms allow companies to adopt their own internal procedures as best befitting them in terms of application and implementation; secondly, the Norms require businesses to evaluate their own major activities in the light of the provisions; third, the Norms require transparency and input from the various stakeholders; fourth, the Norms call for compensation, reparation or restoration in case of violations; and finally, the Norms call on the governments to draw up a framework for the application of the Norms.

The Norms not only involve both governments and companies alike, but they also encourage legislatures to adopt them as part of domestic law. However, they have been questioned, particularly by the International Chamber of Commerce (ICC) and the International Organization of Employers (IOE) on the basis that it is questionable as to whether companies, as non-State actors, can be brought under human rights standards. This query goes counter to the work done by the United Nations as well as the fundamental principle the Universal Declaration of Human Rights cited at the commencement of this article of. As a compromise, the ICC and IOE have indicated that the Norms will be acceptable only as voluntary guidelines.

In essence the entire process of ensuring human rights, and the environment in which they are applied lies in governance. The concept of "governance" is as old as human civilization. The most simplistic definition of "governance" would be that it is the process of decision-making and the process by which decisions are implemented (or not implemented as the case may be). Governance can be categorized into several institutional bases and used in several contexts such as corporate governance, international governance, national governance and local governance.

The durability of tort law lies in its enduring ability to remedies to new civil wrongs. New approaches to areas such as medical malpractice and the rights of the patient are constantly evolving to the extent that in the United States, patients rights are being increasingly enforced against medical malpractice or negligence of physicians or medical institutions such as hospitals and clinics. The fundamental principle of professional conduct obtaining in the United States is that the physician is in a position of trust and confidence as regards the patient, and it is her duty to act with the utmost good faith toward the patient. If she knows that she cannot affect a cure, or that the treatment adopted will probably be of no benefit, it is her duty to advise the patient of these facts.[9] English law has accepted the concept of *incrementalism*,[10] which rejects generalization in relation to the duty of care, in favor of a cautious development of law founded on analogies to similar fact situations, but espousing and applying *fairness and justice* to each case. This approach has its genesis in the judgment of Justice Brennan in the 1985 case of *Sutherland Shire Council* v. *Heyman*[11] handed down by the High Court of Australia where His Honour said:

> It is preferable. . .that the law should develop novel categories of negligence incrementally and by analogy with existing categories, rather than by massive extension of a prima facie duty of care restrained only by the indefinable. . .considerations which ought to negative, or to reduce or limit the scope of the duty or the class of person to whom it is owed.[12]

The House of Lords found it fit to import this approach to the United Kingdom in the leading 1990 case of *Caparo Industries Plc* v. *Dickman*[13] where Lord Bridge stated:

> Whilst recognizing, of course, the importance of the underlying general principles common to the whole field of negligence, I think the law has now moved in the direction of attaching greater significance to the more traditional categorization of distinct and recognizable situation, as guides to the existence, the scope and the limits of the varied duties of care which the law imposes.[14]

Therefore, there is no simple formula or touchstone to which recourse can be had in order to provide in every case a ready answer to questions as to whether the law will or will not grant recourse based on traditional rules of negligence. In this context, one wonders whether the use of such catch phrases as "reasonably skilled professional," without any attendant criteria to define the phrase is practical anymore. It would certainly be interesting if the "incrementalism" approach were to be applied along with established rules of law to future instances of adjudication on negligence.

Finally, a disconcerting fact is that it is claimed that the main obstacle to tort reform is that although it is desirable to keep an ongoing process of tort reform, those involved in such reform are largely duplicitous in claiming to be grass roots

[9]See Regan, *Doctor, Patient and the Law* (1950) at 34. See also, *Benson* v. *Dean* 232 NY 52; 133 NE 125 (1921). Stryker, *Courts and Doctors* at page 9 says: "The relationship of patient and physician is to the highest possible degree a fiduciary one, involving every element of trust and confidence." The American Medical Association, *Principles of Medical Ethics*, in para. 8 provides: "a physician should seek consultation upon request, in doubtful or difficult cases, or whenever it appears that the quality of medical service may be enhanced thereby." See also, Annot, Duty to Send Patient to Specialist, 132 *ALR* 392 (1949).

[10]See Stanton (1997).

[11](1985) 157 *CLR* 424.

[12](1985) 157 *CLR* 481.

[13][1990] 2 *AC* 605.

[14][1990] 2 *AC* 618.

citizens alleging to speak for the average American, while their tax filings show that they are among the top level of tax payers and occupy high corporate positions having largely successful industrial profiles. Often the industries they represent are the defenders in cases ranging from defective products to securities scams and medical malpractice.

References

Dudley J, Baylies E (1876) Addison on torts, American edn. C.C. Soule, Boston
Osborne PH (2003) The law of torts, 2nd edn. Irwin Law, Toronto, ON
Prosser WL (1964) Handbook of the law of torts, 3rd edn. West Publishing Company, St. Paul
Stanton K (1997) Incremental approaches to the duty of care, Chap 2. In: Mullany N (ed) Torts in the nineties. Law Book Company
White GE (1980) Tort law in America. Oxford University Press, Oxford

Appendix

SPACE LAW TREATIES

THE OUTER SPACE TREATY

TREATY ON PRINCIPLES GOVERNING THE ACTIVITIES OF STATES IN THE EXPLORATION AND USE OF OUTER SPACE, INCLUDING THE MOON AND OTHER CELESTIAL BODIES

Signed at Washington, London, Moscow, January 27, 1967
Ratification advised by U.S. Senate April 25, 1967
Ratified by U.S. President May 24, 1967
U.S. ratification deposited at Washington, London, and Moscow October 10, 1967
Proclaimed by U.S. President October 10, 1967
Entered into force October 10, 1967

The States Parties to this Treaty,

Inspired by the great prospects opening up before mankind as a result of mans entry into outer space,

Recognizing the common interest of all mankind in the progress of the exploration and use of outer space for peaceful purposes,

Believing that the exploration and use of outer space should be carried on for the benefit of all peoples irrespective of the degree of their economic or scientific development,

Desiring to contribute to broad international co-operation in the scientific as well as the legal aspects of the exploration and use of outer space for peaceful purposes,

Believing that such co-operation will contribute to the development of mutual understanding and to the strengthening of friendly relations between States and peoples,

Recalling resolution 1962 (XVIII), entitled "Declaration of Legal Principles Governing the Activities of States in the Exploration and Use of Outer Space", which was adopted unanimously by the United Nations General Assembly on 13 December 1963,

R. Abeyratne, *Space Security Law*,
DOI 10.1007/978-3-642-16702-7, © Springer-Verlag Berlin Heidelberg 2011

Recalling resolution 1884 (XVIII), calling upon States to refrain from placing in orbit around the Earth any objects carrying nuclear weapons or any other kinds of weapons of mass destruction or from installing such weapons on celestial bodies, which was adopted unanimously by the United Nations General Assembly on 17 October 1963,

Taking account of United Nations General Assembly resolution 110 (II) of 3 November 1947, which condemned propaganda designed or likely to provoke or encourage any threat to the peace, breach of the peace or act of aggression, and considering that the aforementioned resolution is applicable to outer space,

Convinced that a Treaty on Principles Governing the Activities of States in the Exploration and Use of Outer Space, including the Moon and Other Celestial Bodies, will further the Purposes and Principles of the Charter of the United Nations,

Have agreed on the following:

Article I

The exploration and use of outer space, including the moon and other celestial bodies, shall be carried out for the benefit and in the interests of all countries, irrespective of their degree of economic or scientific development, and shall be the province of all mankind.

Outer space, including the moon and other celestial bodies, shall be free for exploration and use by all States without discrimination of any kind, on a basis of equality and in accordance with international law, and there shall be free access to all areas of celestial bodies.

There shall be freedom of scientific investigation in outer space, including the moon and other celestial bodies, and States shall facilitate and encourage international co-operation in such investigation.

Article II

Outer space, including the moon and other celestial bodies, is not subject to national appropriation by claim of sovereignty, by means of use or occupation, or by any other means.

Article III

States Parties to the Treaty shall carry on activities in the exploration and use of outer space, including the moon and other celestial bodies, in accordance with international law, including the Charter of the United Nations, in the interest of maintaining international peace and security and promoting international co-operation and understanding.

Article IV

States Parties to the Treaty undertake not to place in orbit around the Earth any objects carrying nuclear weapons or any other kinds of weapons of mass destruction, install such weapons on celestial bodies, or station such weapons in outer space in any other manner.

The Moon and other celestial bodies shall be used by all States Parties to the Treaty exclusively for peaceful purposes. The establishment of military bases, installations and fortifications, the testing of any type of weapons and the conduct of military maneuvers on celestial bodies shall be forbidden. The use of military personnel for scientific research or for any other peaceful purposes shall not be prohibited. The use of any equipment or facility necessary for peaceful exploration of the Moon and other celestial bodies shall also not be prohibited.

Article V

States Parties to the Treaty shall regard astronauts as envoys of mankind in outer space and shall render to them all possible assistance in the event of accident, distress, or emergency landing on the territory of another State Party or on the high seas. When astronauts make such a landing, they shall be safely and promptly returned to the State of registry of their space vehicle.

In carrying on activities in outer space and on celestial bodies, the astronauts of one State Party shall render all possible assistance to the astronauts of other States Parties.

States Parties to the Treaty shall immediately inform the other States Parties to the Treaty or the Secretary-General of the United Nations of any phenomena they discover in outer space, including the Moon and other celestial bodies, which could constitute a danger to the life or health of astronauts.

Article VI

States Parties to the Treaty shall bear international responsibility for national activities in outer space, including the Moon and other celestial bodies, whether such activities are carried on by governmental agencies or by non-governmental entities, and for assuring that national activities are carried out in conformity with the provisions set forth in the present Treaty. The activities of non-governmental entities in outer space, including the Moon and other celestial bodies, shall require authorization and continuing supervision by the appropriate State Party to the Treaty. When activities are carried on in outer space, including the Moon and other celestial bodies, by an international organization, responsibility for compliance with this Treaty shall be borne both by the international organization and by the States Parties to the Treaty participating in such organization.

Article VII

Each State Party to the Treaty that launches or procures the launching of an object into outer space, including the Moon and other celestial bodies, and each State Party from whose territory or facility an object is launched, is internationally liable for damage to another State Party to the Treaty or to its natural or juridical persons by such object or its component parts on the Earth, in air space or in outer space, including the Moon and other celestial bodies.

Article VIII

A State Party to the Treaty on whose registry an object launched into outer space is carried shall retain jurisdiction and control over such object, and over any personnel thereof, while in outer space or on a celestial body. Ownership of objects launched into outer space, including objects landed or constructed on a celestial body, and of their component parts, is not affected by their presence in outer space or on a celestial body or by their return to the Earth. Such objects or component parts found beyond the limits of the State Party to the Treaty on whose registry they are carried shall be returned to that State Party, which shall, upon request, furnish identifying data prior to their return.

Article IX

In the exploration and use of outer space, including the Moon and other celestial bodies, States Parties to the Treaty shall be guided by the principle of co-operation and mutual assistance and shall conduct all their activities in outer space, including the Moon and other celestial bodies, with due regard to the corresponding interests of all other States Parties to the Treaty. States Parties to the Treaty shall pursue studies of outer space, including the Moon and other celestial bodies, and conduct exploration of them so as to avoid their harmful contamination and also adverse changes in the environment of the Earth resulting from the introduction of extraterrestrial matter and, where necessary, shall adopt appropriate measures for this purpose. If a State Party to the Treaty has reason to believe that an activity or experiment planned by it or its nationals in outer space, including the

Moon and other celestial bodies, would cause potentially harmful interference with activities of other States Parties in the peaceful exploration and use of outer space, including the Moon and other celestial bodies, it shall undertake appropriate international consultations before proceeding with any such activity or experiment. A State Party to the Treaty which has reason to believe that an activity or experiment planned by another State Party in outer space, including the Moon and other celestial bodies, would cause potentially harmful interference with activities in the peaceful exploration and use of outer space, including the Moon and other celestial bodies, may request consultation concerning the activity or experiment.

Article X

In order to promote international co-operation in the exploration and use of outer space, including the Moon and other celestial bodies, in conformity with the purposes of this Treaty, the States Parties to the Treaty shall consider on a basis of equality any requests by other States Parties to the Treaty to be afforded an opportunity to observe the flight of space objects launched by those States.

The nature of such an opportunity for observation and the conditions under which it could be afforded shall be determined by agreement between the States concerned.

Article XI

In order to promote international co-operation in the peaceful exploration and use of outer space, States Parties to the Treaty conducting activities in outer space, including the Moon and other celestial bodies, agree to inform the Secretary-General of the United Nations as well as the public and the international scientific community, to the greatest extent feasible and practicable, of the nature, conduct, locations and results of such activities. On receiving the said information, the Secretary-General of the United Nations should be prepared to disseminate it immediately and effectively.

Article XII

All stations, installations, equipment and space vehicles on the Moon and other celestial bodies shall be open to representatives of other States Parties to the Treaty on a basis of reciprocity. Such representatives shall give reasonable advance notice of a projected visit, in order that appropriate consultations may be held and that maximum precautions may be taken to assure safety and to avoid interference with normal operations in the facility to be visited.

Article XIII

The provisions of this Treaty shall apply to the activities of States Parties to the Treaty in the exploration and use of outer space, including the Moon and other celestial bodies, whether such activities are carried on by a single State Party to the Treaty or jointly with other States, including cases where they are carried on within the framework of international intergovernmental organizations.

Any practical questions arising in connection with activities carried on by international inter-governmental organizations in the exploration and use of outer space, including the Moon and other celestial bodies, shall be resolved by the States Parties to the Treaty either with the appropriate international organization or with one or more States members of that international organization, which are Parties to this Treaty.

Article XIV

1. This Treaty shall be open to all States for signature. Any State which does not sign this Treaty before its entry into force in accordance with paragraph 3 of this article may accede to it at any time.
2. This Treaty shall be subject to ratification by signatory States. Instruments of ratification and instruments of accession shall be deposited with the Governments of the United States of America, the United Kingdom of Great Britain and Northern Ireland and the Union of Soviet Socialist Republics, which are hereby designated the Depositary Governments.
3. This Treaty shall enter into force upon the deposit of instruments of ratification by five Governments including the Governments designated as Depositary Governments under this Treaty.
4. For States whose instruments of ratification or accession are deposited subsequent to the entry into force of this Treaty, it shall enter into force on the date of the deposit of their instruments of ratification or accession.
5. The Depositary Governments shall promptly inform all signatory and acceding States of the date of each signature, the date of deposit of each instrument of ratification of and accession to this Treaty, the date of its entry into force and other notices.
6. This Treaty shall be registered by the Depositary Governments pursuant to Article 102 of the Charter of the United Nations.

Article XV

Any State Party to the Treaty may propose amendments to this Treaty. Amendments shall enter into force for each State Party to the Treaty accepting the amendments upon their acceptance by a majority of the States Parties to the Treaty and thereafter for each remaining State Party to the Treaty on the date of acceptance by it.

Article XVI

Any State Party to the Treaty may give notice of its withdrawal from the Treaty one year after its entry into force by written notification to the Depositary Governments. Such withdrawal shall take effect one year from the date of receipt of this notification.

Article XVII

This Treaty, of which the English, Russian, French, Spanish and Chinese texts are equally authentic, shall be deposited in the archives of the Depositary Governments. Duly certified copies of this Treaty shall be transmitted by the Depositary Governments to the Governments of the signatory and acceding States.

IN WITNESS WHEREOF the undersigned, duly authorized, have signed this Treaty.

DONE in triplicate, at the cities of Washington, London and Moscow, this twenty-seventh day of January one thousand nine hundred sixty-seven.

Source: Department of State

THE MOON TREATY

ENTERED INTO FORCE: 11 July 1984

The States Parties to this Agreement,

Noting the achievements of States in the exploration and use of the moon and other celestial bodies,

Recognizing that the moon, as a natural satellite of the earth, has an important role to play in the exploration of outer space,

Determined to promote on the basis of equality the further development of co-operation among States in the exploration and use of the moon and other celestial bodies,

Desiring to prevent the moon from becoming an area of international conflict,

Bearing in mind the benefits which may be derived from the exploitation of the natural resources of the moon and other celestial bodies,

Recalling the Treaty on Principles Governing the Activities of States in the Exploration and Use of Outer Space, including the Moon and Other Celestial Bodies, the Agreement on the Rescue of Astronauts, the Return of Astronauts and the Return of Objects Launched into Outer Space, the Convention on International Liability for Damage Caused by Space Objects, and the Convention on Registration of Objects Launched into Outer Space,

Taking into account the need to define and develop the provisions of these international instruments in relation to the moon and other celestial bodies, having regard to further progress in the exploration and use of outer space,

Have agreed on the following:

Article 1

1. The provisions of this Agreement relating to the moon shall also apply to other celestial bodies within the solar system, other than the earth, except in so far as specific legal norms enter into force with respect to any of these celestial bodies.
2. For the purposes of this Agreement reference to the moon shall include orbits around or other trajectories to or around it.
3. This Agreement does not apply to extraterrestrial materials which reach the surface of the earth by natural means.

Article 2

All activities on the moon, including its exploration and use, shall be carried out in accordance with international law, in particular the Charter of the United Nations, and taking into account the Declaration on Principles of International Law concerning Friendly Relations and Co-operation Among States in accordance with the Charter of the United Nations, adopted by the General Assembly on 24 October 1970, in the interests of maintaining international peace and security and promoting international co-operation and mutual understanding, and with due regard to the corresponding interests of all other States Parties.

Article 3

1. The moon shall be used by all States Parties exclusively for peaceful purposes.
2. Any threat or use of force or any other hostile act or threat of hostile act on the moon is prohibited. It is likewise prohibited to use the moon in order to commit any such act or to engage in any such threat in relation to the earth, the moon, spacecraft, the personnel of spacecraft or man-made space objects.
3. States Parties shall not place in orbit around or other trajectory to or around the moon objects carrying nuclear weapons or any other kinds of weapons of mass destruction or place or use such weapons on or in the moon.
4. The establishment of military bases, installations and fortifications, the testing of any type of weapons and the conduct of military manoeuvres on the moon shall be forbidden. The use of military personnel for scientific research or for any other peaceful purposes shall not be prohibited. The use of any equipment or facility necessary for peaceful exploration and use of the moon shall also not be prohibited.

Article 4

1. The exploration and use of the moon shall be the province of all mankind and shall be carried out for the benefit and in the interests of all countries, irrespective of their degree of economic or scientific development. Due regard shall be paid to the interests of present and future generations as well as to the need to promote higher standards of living and conditions of economic and social progress and development in accordance with the Charter of the United Nations.
2. States Parties shall be guided by the principle of co-operation and mutual assistance in all their activities concerning the exploration and use of the moon. International co-operation in pursuance of this Agreement should be as wide as possible and may take place on a multilateral basis, on a bilateral basis or through international intergovernmental organizations.

Article 5

1. States Parties shall inform the Secretary-General of the United Nations as well as the public and the international scientific community, to the greatest extent feasible and practicable, of their activities concerned with the exploration and use of the moon. Information on the time, purposes, locations, orbital parameters and duration shall be given in respect of each mission to the moon as soon as possible after launching, while information on the results of each mission, including scientific results, shallbe furnished upon completion of the mission. In the case of a mission lasting more than thirty days, information on conduct of the mission, including any scientific results, shall be given periodically at thirty days' intervals. For missions lasting more than six months, only significant additions to such information need be reported thereafter.
2. If a State Party becomes aware that another State Party plans to operate simultaneously in the same area of or in the same orbit around or trajectory to or around the moon, it shall promptly inform the other State of the timing of and plans for its own operations.
3. In carrying out activities under this Agreement, States Parties shall promptly inform the Secretary-General, as well as the public and the international scientific community, of any phenomena they discover in outer space, including the moon, which could endanger human life or health, as well as of any indiation of organic life.

Article 6

1. There shall be freedom of scientific investigation on the moon by all States Parties without discrimination of any kind, on the basis of equality and in accordance with international law.
2. In carrying out scientific investigations and in furtherance of the provisions of this Agreement, the States Parties shall have the right to collect on and remove from the moon samples of its mineral and other substances. Such samples shall remain at the disposal of those States Parties which caused them to be collected and may be used by them for scientific purposes. States Parties shall have regard to the desirability of making a portion of such samples available to other interested States Parties and the international scientific community for scientific investigation. States Parties may in the course of scientific investigations also use mineral and other substances of the moon in quantities appropriate for the support of their missions.
3. States Parties agree on the desirability of exchanging scientific and other personnel on expeditions to or installations on the moon to the greatest extent feasible and practicable.

Article 7

1. In exploring and using the moon, States Parties shall take measures to prevent the disruption of the existing balance of its environment whether by introducing adverse changes in that environment, by its harmful contamination through the introduction of extra-environmental matter or otherwise. States Parties shall also take measures to avoid harmfully affecting the environment of the earth through the introduction of extraterrestrial matter or otherwise.
2. States Parties shall inform the Secretary-General of the United Nations of the measures being adopted by them in accordance with paragraph 1 of this article and shall also, to the maximum extent feasible, notify him in advance of all placements by them of radio-active materials on the moon and of the purposes of such placements.
3. States Parties shall report to other States Parties and to the Secretary-General concerning areas of the moon having special scientific interest in order that, without prejudice to the rights of other States Parties, consideration may be given to the designation of such areas as international scientific preserves for which special protective arrangements are to be agreed upon in consultation with the competent bodies of the United Nations.

Article 8

1. States Parties may pursue their activities in the exploration and use of the moon anywhere on or below its surface, subject to the provisions of this Agreement.
2. For these purposes States Parties may, in particular:
   (a) Land their space objects on the moon and launch them from the moon;
   (b) Place their personnel, space vehicles, equipment, facilities, stations and installations anywhere on or below the surface of the moon.

   Personnel, space vehicles, equipment, facilities, stations and installations may move or be moved freely over or below the surface of the moon.
3. Activities of States Parties in accordance with paragraphs 1 and 2 of this article shall not interfere with the activities of other States Parties on the moon. Where such interference may occur, the States Parties concerned shall undertake consultations in accordance with article 15, paragraphs 2 and 3 of this Agreement.

Article 9

1. States Parties may establish manned and unmanned stations on the moon. A State Party establishing a station shall use only that area which is required

for the needs of the station and shall immediately inform the Secretary-General of the United Nations of the location and purposes of that station. Subsequently, at annual intervals that State shall likewise inform the Secretary-General whether the station continues in use and whether its purposes have changed.

2. Stations shall be installed in such a manner that they do not impede the free access to all areas of the moon by personnel, vehicles and equipment of other States Parties conducting activities on the moon in accordance with the provisions of this Agreement or of article I of the Treaty on Principles Governing the Activities of States in the Exploration and Use of Outer Space, including the Moon and Other Celestial Bodies.

Article 10

1. States Parties shall adopt all practicable measures to safeguard the life and health of persons on the moon. For this purpose they shall regard any person on the moon as an astronaut within the meaning of article V of the Treaty on Principles Governing the Activities of States in the Exploration and Use of Outer Space, including the Moon and Other Celestial Bodies and as part of the personnel of a spacecraft within the meaning of the Agreement on the Rescue of Astronauts, the Return of Astronauts and the Return of Objects Launched into Outer Space.

2. States Parties shall offer shelter in their stations, installations, vehicles and other facilities to persons in distress on the moon.

Article 11

1. The moon and its natural resources are the common heritage of mankind, which finds its expression in the provisions of this Agreement and in particular in paragraph 5 or this article.

2. The moon is not subject to national appropriation by any claim of sovereignty, by means of use or occupation, or by any other means.

3. Neither the surface nor the subsurface of the moon, nor any part thereof or natural resources in place, shall become property of any State, international intergovernmental or non-governmental organization, national organization or non-governmental entity or of any natural person. The placement of personnel, space vehicles, equipment, facilities, stations and installations on or below the surface of the moon, including structures connected with its surface or subsurface, shall not create a right of ownership over the surface or the subsurface of the moon or any areas thereof. The foregoing provisions are without prejudice to the international regime referred to in paragraph 5 of this article.

4. States Parties have the right to exploration and use of the moon without discrimination of any kind, on a basis of equality and in accordance with international law and the terms of this Agreement.

5. States Parties to this Agreement hereby undertake to establish an international regime, including appropriate procedures, to govern the exploitation of the natural resources of the moon as such exploitation is about to become feasible. This provision shall be implemented in accordance with article 18 of this Agreement.

6. In order to facilitate the establishment of the international regime referred to in paragraph 5 of this article, States Parties shall inform the Secretary-General of the United Nations as well as the public and the international scientific community, to the greatest extent feasible and practicable, of any natural resources they may discover on the moon.

7. The main purposes of the international regime to be established shall include:

    (a) The orderly and safe development of the natural resources of the moon;

    (b) The rational management of those resources;

    (c) The expansion of opportunities in the use of those resources;

    (d) An equitable sharing by all States Parties in the benefits derived from those resources, whereby the interests and needs of the developing countries, as well as the efforts of those countries which have contributed either directly or indirectly to the exploration of the moon, shall be given special consideration.

8. All the activities with respect to the natural resources of the moon shall be carried out in a manner compatible with the purposes specified in paragraph 7 of this article and the provisions of article 6, paragraph 2, of this Agreement.

Article 12

1. States Parties shall retain jurisdiction and control over their personnel, vehicles, equipment, facilities, stations and installations on the moon. The ownership of space vehicles, equipment, facilities, stations and installations shall not be affected by their presence on the moon.

2. Vehicles, installations and equipment or their component parts found in places other than their intended location shall be dealt with in accordance with article 5 of the Agreement on Rescue of Astronauts, the Return of Astronauts and the Return of Objects Launched into Outer Space.

3. In the event of an emergency involving a threat to human life, States Parties may use the equipment, vehicles, installations, facilities or supplies of other States Parties on the moon. Prompt notification of such use shall be made to the Secretary-General of the United Nations or the State Party concerned.

Article 13

A State Party which learns of the crash landing, forced landing or other unintended landing on the moon of a space object, or its component parts, that were not launched by it, shall promptly inform the launching State Party and the Secretary-General of the United Nations.

Article 14

1. States Parties to this Agreement shall bear international responsibility for national activities on the moon, whether such activities are carried on by governmental agencies or by non-governmental entities, and for assuring that national activities are carried out in conformity with the provisions set forth in this Agreement. States Parties shall ensure that non-governmental entities under their jurisdiction shall engage in activities on the moon only under the authority and continuing supervision of the appropriate State Party.
2. States Parties recognize that detailed arrangements concerning liability for damage caused on the moon, in addition to the provisions of the Treaty on Principles Governing the Activities of States in the Exploration and Use of Outer Space, including the Moon and Other Celestial Bodies and the Convention on International Liability for Damage Caused by Space Objects, may become necessary as a result of more extensive activities on the moon. Any such arrangements shall be elaborated in accordance with the procedure provided for in article 18 of this Agreement.

Article 15

1. Each State Party may assure itself that the activities of other States Parties in the exploration and use of the moon are compatible with the provisions of this Agreement. To this end, all space vehicles, equipment, facilities, stations and installations on the moon shall be open to other States Parties. Such States Parties shall give reasonable advance notice of a projected visit, in order that appropriate consultations may be held and that maximum precautions may be taken to assure safety and to avoid interference with normal operations in the facility to be visited. In pursuance of this article, any State Party may act on its own behalf or with the full or partial assistance of any other State Party or through appropriate international procedures within the framework of the United Nations and in accordance with the Charter.
2. A State Party which has reason to believe that another State Party is not fulfilling the obligations incumbent upon it pursuant to this Agreement or that another State Party is interfering with the rights which the former State has under this Agreement may request consultations with that State Party. A State Party receiving such a request shall enter into such consultations without delay.

Any other State Party which requests to do so shall be entitled to take part in the consultations. Each State Party participating in such consultations shall seek a mutually acceptable resolution of any controversy and shall bear in mind the rights and interests of all States Parties. The Secretary-General of the United Nations shall be informed of the results of the consultations and shall transmit the information received to all States Parties concerned.

3. If the consultations do not lead to a mutually acceptable settlement which has due regard for the rights and interests of all States Parties, the parties concerned shall take all measures to settle the dispute by other peaceful means of their choice appropriate to the circumstances and the nature of the dispute. If difficulties arise in connexion with the opening of consultations or if consultations do not lead to a mutually acceptable settlement, any State Party may seek the assistance of the Secretary-General, without seeking the consent of any other State Party concerned, in order to resolve the controversy. A State Party which does not maintain diplomatic relations with another State Party concerned shall participate in such consultations, at its choice, either itself or through another State Party or the Secretary-General as intermediary.

Article 16

With the exception of articles 17 to 21, references in this Agreement to States shall be deemed to apply to any international intergovernmental organization which conducts space activities if the organization declares its acceptance of the rights and obligations provided for in this Agreement and if a majority of the States members of the organization are States Parties to this Agreement and to the Treaty on Principles Governing the Activities of States in the Exploration and Use of Outer Space, including the Moon and Other Celestial Bodies. States members of any such organization which are States Parties to this Agreement shall take all appropriate steps to ensure that the organization makes a declaration in accordance with the foregoing.

Article 17

Any State Party to this Agreement may propose amendments to the Agreement. Amendments shall enter into force for each State Party to the Agreement accepting the amendments upon their acceptance by a majority of the States Parties to the Agreement and thereafter for each remaining State Party to the Agreement on the date of acceptance by it.

Article 18

Ten years after the entry into force of this Agreement, the question of the review of the Agreement shall be included in the provisional agenda of the

General Assembly of the United Nations in order to consider, in the light of past application of the Agreement, whether it requires revision. However, at any time after the Agreement has been in force for five years, the Secretary-General of the United Nations, as depository, shall, at the request of one third of the States Parties to the Agreement and with the concurrence of the majority of the States Parties, convene a conference of the States Parties to review this Agreement. A review conference shall also consider the question of the implementation of the provisions of article 11, paragraph 5, on the basis of the principle referred to in paragraph 1 of that article and taking into account in particular any relevant technological developments.

Article 19

1. This Agreement shall be open for signature by all States at United Nations Headquarters in New York.
2. This Agreement shall be subject to ratification by signatory States. Any State which does not sign this Agreement before its entry into force in accordance with paragraph 3 of this article may accede to it at any time. Instruments of ratification or accession shall be deposited with the Secretary-General of the United Nations.
3. This Agreement shall enter into force on the thirtieth day following the date of deposit of the fifth instrument of ratification.
4. For each State depositing its instrument of ratification or accession after the entry into force of this Agreement, it shall enter into force on the thirtieth day following the date of deposit of any such instrument.
5. The Secretary-General shall promptly inform all signatory and acceding States of the date of each signature, the date of deposit of each instrument of ratification or accession to this Agreement, the date of its entry into force and other notices.

Article 20

Any State Party to this Agreement may give notice of its withdrawal from the Agreement one year after its entry into force by written notification to the Secretary-General of the United Nations. Such withdrawal shall take effect one year from the date of receipt of this notification.

Article 21

The original of this Agreement, of which the Arabic, Chinese, English, French, Russian and Spanish texts are equally authentic, shall be deposited with the

Secretary-General of the United Nations, who shall send certified copies thereof to all signatory and acceding States.

In witness whereof the undersigned, being duly authorized thereto by their respective Governments, have signed this Agreement, opened for signature at New York on December 18, 1979.

THE LIABILITY CONVENTION

CONVENTION

ON INTERNATIONAL LIABILITY

FOR DAMAGE CAUSED

BY SPACE OBJECTS

(1972)

ENTERED INTO FORCE: 1 September 1972

The States Parties to this Convention,

Recognising the common interest of all mankind in furthering the exploration and use of outer space for peaceful purposes,

Recalling the Treaty on Principles Governing the Activities of States in the Exploration and Use of Outer Space, including the Moon and Other Celestial Bodies,

Taking into consideration that, notwithstanding the precautionary measures to be taken by States and international intergovernmental organisations involved in the launching of space objects, damage may on occasion be caused by such objects,

Recognizing the need to elaborate effective international rules and procedures concerning liability for damage caused by space objects and to ensure, in particular, the prompt payment under the terms of this Convention of a full and equitable measure of compensation to victims of such damage,

Believing that the establishment of such rules and procedures will contribute to the strengthening of international co-operation in the field of the exploration and use of outer space for peaceful purposes,

Have agreed on the following:

Article I. For the purposes of this Convention:

(a) The term "damage" means loss of life, personal injury or other impairment of health; or loss of or damage to property of States or of persons, natural or juridical, or property of international intergovernmental organisations;

(b) The term "launching" includes attempted launching;

(c) The term "launching State" means:

(i). a state which launches or procures the launching of a space object;
(ii). a State from whose territory or facility a space object is launched;

(d) The term "space object" includes component parts of a space object as well as its launch vehicle and parts thereof.

Article II. A launching State shall be absolutely liable to pay compensation for damage caused by its space object on the surface of the earth or to aircraft in flight.

Article III. In the event of damage being caused elsewhere than on the surface of the earth to a space object of one launching State or to persons or property on board such a space object by a space object of another launching State, the latter shall be liable only if the damage is due to its fault or the fault of persons for whom it is responsible.

Article IV.

1. In the event of damage being caused elsewhere than on the surface of the earth to a space object of one launching State or to persons or property on board such a space object by a space object of another launching State, and of damage thereby being caused to a third State or to its natural or juridical persons, the first two States shall be jointly and severally liable to the third State, to the extent indicated by the following:

   (a) If the damage has been caused to the third State on the surface of the earth or to aircraft in flight, their liability to the third State shall be absolute;

   (b) If the damage has been caused to a space object of the third State or to persons or property on board that space object elsewhere than on the surface of the earth, their liability to the third State shall be based on the fault of either of the first two States or on the fault of persons for whom either is responsible.

2. In all cases of joint and several liability referred to in paragraph 1 of this Article, the burden of compensation for the damage shall be apportioned between the first two States in accordance with the extent to which they were at fault; if the extent of the fault of each of these States cannot be established, the burden of compensation shall be apportioned equally between them. Such apportionment shall be without prejudice to the right of the third State to seek the entire compensation due under this Convention from any or all of the launching States which are jointly and severally liable.

Article V.

1. Whenever two or more States jointly launch a space object, they shall be jointly and severally liable for any damage caused.

2. A launching State which has paid compensation for damage shall have the right to present a claim for indemnification to other participants in the joint launching. The participants in a joint launching may conclude agreements regarding the apportioning among themselves of the financial obligation in respect of which they are jointly and severally liable. Such agreements shall be without prejudice to the right of a State sustaining damage to seek the entire compensation due

under this Convention from any or all of the launching States which are jointly and severally liable.

3. A State from whose territory or facility a space object is launched shall be regarded as a participant in a joint launching.

Article VI.

1. Subject to the provisions of paragraph 2 of this Article, exoneration from absolute liability shall be granted to the extent that a launching State establishes that the damage has resulted either wholly or partially from gross negligence or from an act or omission done with intent to cause damage on the part of a claimant State or of natural or juridical persons it represents.

2. No exoneration whatever shall be granted in cases where the damage has resulted from activities conducted by a launching State which are not in conformity with international law including, in particular, the Charter of the United Nations and the Treaty on Principles Governing the Activities of States in the Exploration and Use of Outer Space, including the Moon and Other Celestial Bodies.

Article VII. The provisions of this Convention shall not apply to damage caused by a space object of a launching State to:

(a) nationals of that launching State;

(b) foreign nationals during such time as they are participating in the operation of that space object from the time of its launching or at any stage thereafter until its descent, or during such time as they are in the immediate vicinity of a planned launching or recovery area as the result of an invitation by that launching State.

Article VIII.

1. A State which suffers damage, or whose natural or juridical persons suffer damage, may present to a launching State a claim for compensation for such damage.

2. If the State of nationality has not presented a claim, another State may, in respect of damage sustained in its territory by any natural or juridical person, present a claim to a launching State.

3. If neither the State of nationality nor the State in whose territory the damage was sustained has presented a claim or notified its intention of presenting a claim, another State may, in respect of damage sustained by its permanent residents, present a claim to a launching State.

Article IX. A claim for compensation for damage shall be presented to a launching State through diplomatic channels. If a State does not maintain diplomatic relations with the launching State concerned, it may request another State to present its claim to that launching State or otherwise represent its interests under this Convention. It may also present its claim through the Secretary-General of the United Nations, provided the claimant State and the launching State are both Members of the United Nations.

Article X.

1. A claim for compensation for damage may be presented to a launching State not later than one year following the date of the occurrence of the damage or the identification of the launching State which is liable.
2. If, however, a State does not know of the occurrence of the damage or has not been able to identify the launching State which is liable, it ay present a claim within one year following the date on which it learned of the aforementioned facts; however, this period shall in no event exceed one year following the date on which the State could reasonably be expected to have learned of the facts through the exercise of due diligence.
3. The time-limits specified in paragraphs 1 and 2 of this Article shall apply even if the full extent of the damage may not be known. In this event, however, the claimant State shall be entitled to revise the claim and submit additional documentation after the expiration of such time-limits until one year after the full extent of the damage is known.

Article XI.

1. Presentation of a claim to a launching State for compensation for damage under this Convention shall not require the prior exhaustion of any local remedies which may be available to a claimant State or to natural or juridical persons it represents.
2. Nothing in this Convention shall prevent a State, or natural or juridical persons it might represent, from pursuing a claim in the courts or administrative tribunals or agencies of a launching State. A State shall not, however, be entitled to present a claim under this Convention in respect of the same damage for which a claim is being pursued in the courts or administrative tribunals or agencies of a launching State or under another international agreement which is binding on the States concerned.

Article XII. The compensation which the launching State shall be liable to pay for damage under this Convention shall be determined in accordance with international law and the principles of justice and equity, in order to provide such reparation in respect of the damage as will restore the person, natural or juridical, State or international organisation on whose behalf the claim is presented to the condition which would have existed if the damage had not occurred.

Article XIII. Unless the claimant State and the State from which compensation is due under this Convention agree on another form of compensation, the compensation shall be paid in the currency of the claimant State or, if that State so requests, in the currency of the State from which compensation is due.

Article XIV. If no settlement of a claim is arrived at through diplomatic negotiations as provided for in Article IX, within one year from the date on which the claimant State notifies the launching State that it has submitted the documentation of its claim, the parties concerned shall establish a Claims Commission at the request of either party.

Article XV.

1. The Claims Commission shall be composed of three members: one appointed by the claimant State, one appointed by the launching State and the third member, the Chairman, to be chosen by both parties jointly. Each party shall make its appointment within two months of the request for the establishment of the Claims Commission.
2. If no agreement is reached on the choice of the Chairman within four months of the request for the establishment of the Commission, either party may request the Secretary-General of the United Nations to appoint the Chairman within a further period of two months.

Article XVI.

1. If one of the parties does not make its appointment within the stipulated period, the Chairman shall, at the request of the other party, constitute a single-member Claims Commission.
2. Any vacancy which may arise in the Commission for whatever reason shall be filled by the same procedure adopted for the original appointment.
3. The Commission shall determine its own procedure.
4. The Commission shall determine the place or places where it shall sit and all other administrative matters.
5. Except in the case of decisions and awards by a single-member Commission, all decision and awards of the Commission shall be by majority vote.

Article XVII. No increase in the membership of the Claims Commission shall take place by reason of two or more claimant States or launching States being joined in any one proceeding before the Commission. The claimant States so joined shall collectively appoint one member of the Commission in the same manner and subject to the same conditions as would be the case for a single claimant State. When two or more launching States are so joined, they shall collectively appoint one member of the Commission in the same way. If the claimant States or the launching States do not make the appointment within the stipulated period, the Chairman shall constitute a single-member Commission.

Article XVIII. The Claims Commission shall decide the merits of the claim for compensation and determine the amount of compensation payable, if any.

Article XIX.

1. The Claims Commission shall act in accordance with the provisions of Article XII.
2. The decision of the Commission shall be final and binding if the parties have so agreed; otherwise the Commission shall render a final and recommendatory award, which the parties shall consider in good faith. The Commission shall state the reasons for its decision or award.
3. The Commission shall give its decision or award as promptly as possible and no later than one year from the date of its establishment, unless an extension of this period is found necessary by the Commission.
4. The Commission shall make its decision or award public. It shall deliver a certified copy of its decision or award to each of the parties and to the Secretary-General of the United Nations.

Article XX. The expenses in regard to the Claims Commission shall be borne equally by the parties, unless otherwise decided by the Commission.

Article XXI. If the damage caused by a space object presents a large-scale danger to human life or seriously interferes with the living conditions of the population or the functioning of vital centres, the States Parties, and in particular the launching State, shall examine the possibility of rendering appropriate and rapid assistance to the State which has suffered the damage, when it so requests. However, nothing in this Article shall affect the rights or obligations of the States Parties under this Convention.

Article XXII.

1. In this Convention, with the exception of Articles XXIV to XXVII, references to States shall be deemed to apply to any international intergovernmental organisation which conducts space activities if the organisation declares its acceptance of the rights and obligations provided for in this Convention and if a majority of the States members of the organisation are State Parties to this Convention and to the Treaty on Principles Governing the Activities of States in the Exploration and Use of Outer Space, including the Moon and Other Celestial Bodies.
2. States members of any such organisation which are States Parties to this Convention shall take all appropriate steps to ensure that the organisation makes a declaration in accordance with the preceding paragraph.
3. If an international intergovernmental organisation is liable for damage by virtue of the provisions of this Convention, that organisation and those of its members which are States Parties to this Convention shall be jointly and severally liable; provided, however, that:

(a) any claim for compensation in respect of such damage shall be first presented to the organisation;

(b) only where the organisation has not paid, within a period of six months, any sum agreed or determined to be due as compensation for such damage, may the claimant State invoke the liability of the members which are States Parties to this Convention for the payment of that sum.

4. Any claim, pursuant to the provisions of this Convention, for compensation in respect of damage caused to an organisation which has made a declaration in accordance with paragraph 1 of this Article shall be presented by a State member of the organisation which is a State Party to this Convention.

Article XXIII.

1. The provisions of this Convention shall not affect other international agreements in force in so far as relations between the States Parties to such agreements are concerned.
2. No provision of this Convention shall prevent States from concluding international agreements reaffirming, supplementing or extending its provisions.

Article XXIV.

1. This Convention shall be open to all States for signature. Any State which does not sign this Convention before its entry into force in accordance with paragraph 3 of this Article may accede to it at any time.
2. This Convention shall be subject to ratification by signatory States. Instruments of ratification and instruments of accession shall be deposited with the Governments of the United Kingdom of Great Britain and Northern Ireland, the Union of Soviet Socialist Republics and the United States of America, which are hereby designated the Depositary Governments.
3. This Convention shall enter into force on the deposit of the fifth instrument of ratification.
4. For States whose instruments of ratification or accession are deposited subsequent to the entry into force of this Convention, it shall enter into force on the date of the deposit of their instruments of ratification or accession.
5. The Depositary Governments shall promptly inform all signatory and acceding States of the date of each signature, the date of deposit of each instrument of ratification of and accession to this Convention, the date of its entry into force and other notices.
6. This Convention shall be registered by the Depositary Governments pursuant to Article 102 of the Charter of the United Nations.

Article XXV. Any State Party to this Convention may propose amendments to this Convention. Amendments shall enter into force for each State Party to the

Convention accepting the amendments upon their acceptance by a majority of the States Parties to the Convention and thereafter for each remaining State Party on the date of acceptance by it.

Article XXVI. Ten years after the entry into force of this Convention, the question of the review of this Convention shall be included in the provisional agenda of the United Nations General Assembly in order to consider, in the light of past application of the Convention, whether it requires revision. However, at any time after the Convention has been in force for five years, and at the request of one third of the States Parties to the Convention, and with the concurrence of the majority of the States Parties, a conference of the States Parties shall be convened to review this Convention.

Article XXVII. Any State Party to this Convention may give notice of its withdrawal from the Convention one year after its entry into force by written notification to the Depositary Governments. Such withdrawal shall take effect one year from the date of receipt of this notification.

Article XXVIII. This Convention, of which the English, Russian, French, Spanish and Chinese texts are equally authentic, shall be deposited in the archives of the Depositary Governments. Duly certified copies of this Convention shall be transmitted by the Depositary Governments to the Governments of the signatory and acceding States.

In witness whereof the undersigned, duly authorized thereto, have signed this Convention.

Done in triplicate, at the cities of London, Moscow and Washington, this twenty-ninth day of March, one thousand nine hundred and seventy-two.

THE REGISTRATION CONVENTION

CONVENTION
ON REGISTRATION
OF OBJECTS LAUNCHED
INTO OUTER SPACE
(1975)

The States Parties to this Convention,

Recognizing the common interest of all mankind in furthering the exploration and use of outer space for peaceful purposes

Recalling that the Treaty on principles governing the activities of States in the exploration and use of outer space, including the moon and other celestial bodies of 27 January 1967 affirms that States shall bear international responsibility for their national activities in outer space and refers to the State on whose registry an object launched into outer space is carried,

Recalling also that the Agreement on the rescue of astronauts, the return of astronauts and the return of objects launched into outer space of 22 April 1968 provides that a launching authority shall, upon request, furnish identifying data prior to the return of an object it has launched into outer space found beyond the territorial limits of the launching authority,

Recalling further that the Convention on international liability for damage caused by space objects of 29 March 1972 establishes international rules and procedures concerning the liability of launching States for damage caused by their space objects,

Desiring, in the light of the Treaty on principles governing the activities of States in the exploration and use of outer space, including the moon and other celestial bodies, to make provision for the national registration by launching States of space objects launched into outer space,

Desiring further that a central register of objects launched into outer space be established and maintained, on a mandatory basis, by the Secretary-General of the United Nations,

Desiring also to provide for States Parties additional means and procedures to assist in the identification of space objects,

Believing that a mandatory system of registering objects launched into outer space would, in particular, assist in their identification and would contribute to the application and development of international law governing the exploration and use of outer space,

Have agreed on the following:

Article I

For the purposes of this Convention:

(a) The term "launching State" means:

(i). A State which launches or procures the launching of a space object;
(ii). A State from whose territory or facility a space object is launched;

(b) The term "space object" includes component parts of a space object as well as its launch vehicle and parts thereof;

(c) The term "State of registry" means a launching State on whose registry a space object is carried in accordance with article II.

Article II

1. When a space object is launched into earth orbit or beyond, the launching State shall register the space object by means of an entry in an appropriate registry which it shall maintain. Each launching State shall inform the Secretary-General of the United Nations of the establishment of such a registry.
2. Where there are two or more launching States in respect of any such space object, they shall jointly determine which one of them shall register the object in accordance with paragraph 1 of this article, bearing in mind the provisions of article VIII of the Treaty on principles governing the activities of States in the exploration and use of outer space, including the moon and other celestial bodies, and without prejudice to appropriate agreements concluded or to be concluded among the launching States on jurisdiction and control over the space object and over any personnel thereof.
3. The contents of each registry and the conditions under which it is maintained shall be determined by the State of registry concerned.

Article III

1. The Secretary-General of the United Nations shall maintain a Register in which the information furnished in accordance with article IV shall be recorded.
2. There shall be full and open access to the information in this Register.

Article IV

1. Each State of registry shall furnish to the Secretary-General of the United Nations, as soon as practicable, the following information concerning each space object carried on its registry:

   (a) Name of launching State or States;

   (b) An appropriate designator of the space object or its registration number;

   (c) Date and territory or location of launch;

   (d) Basic orbital parameters, including:
      (i). Nodal period,
      (ii). Inclination,
      (iii). Apogee,
      (iv). Perigee;

   (e) General function of the space object.

2. Each State of registry may, from time to time, provide the Secretary- General of the United Nations with additional information concerning a space object carried on its registry.

3. Each State of registry shall notify the Secretary-General of the United Nations, to the greatest extent feasible and as soon as practicable, of space objects concerning which it has previously transmitted information, and which have been but no longer are in earth orbit.

Article V

Whenever a space object launched into earth orbit or beyond is marked with the designator or registration number referred to in article IV, paragraph 1 (b), or both, the State of registry shall notify the Secretary-General of this fact when submitting the information regarding the space object in accordance with article IV. In such case, the Secretary-General of the United Nations shall record this notification in the Register.

Article VI

Where the application of the provisions of this Convention has not enabled a State Party to identify a space object which has caused damage to it or to any of its natural or juridical persons, or which may be of a hazardous or deleterious nature, other States Parties, including in particular States possessing space monitoring and tracking facilities, shall respond to the greatest extent feasible to a request by that State Party, or transmitted through the Secretary-General on its behalf, for assistance under equitable and reasonable conditions in the identification of the object. A State Party making such a request shall, to the greatest extent feasible, submit information as to the time, nature and circumstances of the events giving rise to the request. Arrangements under which such assistance shall be rendered shall be the subject of agreement between the parties concerned.

Article VII

1. In this Convention, with the exception of articles VIII to XII inclusive, references to States shall be deemed to apply to any international intergovernmental organization which conducts space activities if the organization declares its acceptance of the rights and obligations provided for in this Convention and if a majority of the States members of the organization are States Parties to this Convention and to the Treaty on principles governing the activities of States in the exploration and use of outer space, including the moon and other celestial bodies.
2. States members of any such organization which are States Parties to this Convention shall take all appropriate steps to ensure that the organization makes a declaration in accordance with paragraph 1 of this article.

Article VIII

1. This Convention shall be open for signature by all States at United Nations Headquarters in New York. Any State which does not sign this Convention before its entry into force in accordance with paragraph 3 of this article may accede to it at any time.
2. This Convention shall be subject to ratification by signatory States. Instruments of ratification and instruments of accession shall be deposited with the Secretary-General of the United Nations.
3. This Convention shall enter into force among the States which have deposited instruments of ratification on the deposit of the fifth such instrument with the Secretary-General of the United Nations.
4. For States whose instruments of ratification or accession are deposited subsequent to the entry into force of this Convention, it shall enter into force on the date of the deposit of their instruments of ratification or accession.
5. The Secretary-General shall promptly inform all signatory and acceding States of the date of each signature, the date of deposit of each instrument of ratification of and accession to this Convention, the date of its entry into force and other notices.

Article IX

Any State Party to this Convention may propose amendments to the Convention. Amendments shall enter into force for each State Party to the Convention accepting the amendments upon their acceptance by a majority of the States Parties to the Convention and thereafter for each remaining State Party to the Convention on the date of acceptance by it.

Article X

Ten years after the entry into force of this Convention, the question of the review of the Convention shall be included in the provisional agenda of the United Nations General Assembly in order to consider in the light of past application of the Convention, whether it requires revision. However, at any time after the Convention has been in force for five years, at the request of one third of the States Parties to the Convention and with the concurrence of the majority of the States Parties, a conference of the States Parties shall be convened to review this Convention. Such review shall take into account in particular any relevant technological developments, including those relating to the identification of space objects.

Article XI

Any State Party to this Convention may give notice of its withdrawal from the Convention one year after its entry into force by written notification to the Secretary-General of the United Nations. Such withdrawal shall take effect one year from the date of receipt of this notification.

Article XII

The original of this Convention, of which the Arabic, Chinese, English, French, Russian and Spanish texts are equally authentic, shall be deposited with the Secretary-General of the United Nations, who shall send certified copies thereof to all signatory and acceding States.

In witness whereof the undersigned, being duly authorized thereto by their respective Governments, have signed this Convention, opened for signature at New York on the fourteenth day of January one thousand nine hundred and seventy-five.

AGREEMENT ON THE RESCUE OF ASTRONAUTS, THE RETURN OF ASTRONAUTS AND THE RETURN OF OBJECTS LAUNCHED INTO OUTER SPACE (1968)

ENTERED INTO FORCE: 3 December 1968

The Contracting Parties,

Noting the great importance of the Treaty on Principles Governing the Activities of States in the Exploration and Use of Outer Space, including the Moon and Other Celestial Bodies, which calls for the rendering of all possible assistance to astronauts in the event of accident, distress or emergency landing, the prompt and safe return of astronauts, and the return of objects launched into outer space,

Desiring to develop and give further concrete expression to these duties,

Wishing to promote international co-operation in the peaceful exploration and use of outer space,

Prompted by sentiments of humanity,

Have agreed on the following:

Article 1

Each Contracting Party which receives information or discovers that the personnel of a spacecraft have suffered accident or are experiencing conditions of distress or have made an emergency or unintended landing in territory under its jurisdiction or on the high seas or in any other place not under the jurisdiction of any State shall immediately:

(a) notify the launching authority or, if it cannot identify and immediately communicate with the launching authority, immediately make a public announcement by all appropriate means of communication at its disposal;

(b) notify the Secretary-General of the United Nations, who should disseminate the information without delay by all appropriate means of communication at his disposal.

Article 2

If, owing to accident, distress, emergency or unintended landing, the personnel of a spacecraft land in territory under the jurisdiction of a Contracting Party, it shall immediately take all possible steps to rescue them and render them all necessary assistance. It shall inform the launching authority and also the Secretary-General of the United Nations of the steps it is taking and of their progress. If assistance by the launching authority would help to effect a prompt rescue or would contribute substantially to the effectiveness of search and rescue operations, the launching authority shall co-operate with the Contracting Party with a view to the effective conduct of search and rescue operations. Such operations shall be subject to the direction and control of the Contracting Party, which shall act in close and continuing consultation with the launching authority.

Article 3

If information is received or it is discovered that the personnel of a spacecraft have alighted on the high seas or in any other place not under the jurisdiction of any State, those Contracting Parties which are in a position to do so shall, if necessary, extend assistance in search and rescue operations for such personnel to assure their speedy rescue. They shall inform the launching authority and the Secretary-General of the United Nations of the steps they are taking and of their progress.

Article 4

If, owing to accident, distress, emergency or unintended landing, the personnel of a spacecraft land in territory under the jurisdiction of a Contracting Party or have been found on the high seas or in any other place not under the jurisdiction of any State, they shall be safely and promptly returned to representatives of the launching authority.

Article 5

1. Each Contracting Party which receives information or discovers that a space object or its component parts has returned to Earth in territory under its jurisdiction or on the high seas or in any other place not under the jurisdiction of any State, shall notify the launching authority and the Secretary-General of the United Nations.
2. Each Contracting Party having jurisdiction over the territory on which a space object or its component parts has been discovered shall, upon the request of the launching authority and with assistance from that authority if requested, take such steps as it finds practicable to recover the object or component parts.
3. Upon request of the launching authority, objects launched into outer space or their component parts found beyond the territorial limits of the launching authority shall be returned to or held at the disposal of representatives of the

launching authority, which shall, upon request, furnish identifying data prior to their return.

4. Notwithstanding paragraphs 2 and 3 of this Article, a Contracting Party which has reason to believe that a space object or its component parts discovered in territory under its jurisdiction, or recovered by it elsewhere, is of a hazardous or deleterious nature may so notify the launching authority, which shall immediately take effective steps, under the direction and control of the said Contracting Party, to eliminate possible danger of harm.
5. Expenses incurred in fulfilling obligations to recover and return a space object or its component parts under paragraphs 2 and 3 of this Article shall be borne by the launching authority.

Article 6

For the purposes of this Agreement, the term "launching authority" shall refer to the State responsible for launching, or, where an international intergovernmental organization is responsible for launching, that organization, provided that that organization declares its acceptance of the rights and obligations provided for in this Agreement and a majority of the States members of that organization are Contracting Parties to this Agreement and to the Treaty on Principles Governing the Activities of States in the Exploration and Use of Outer Space, including the Moon and Other Celestial Bodies.

Article 7

1. This Agreement shall be open to all States for signature. Any State which does not sign this Agreement before its entry into force in accordance with paragraph 3 of this Article may accede to it at any time.
2. This Agreement shall be subject to ratification by signatory States. Instruments of ratification and instruments of accession shall be deposited with the Governments of the United Kingdom of Great Britain and Northern Ireland, the Union of Soviet Socialist Republics and the United States of America, which are hereby designated the Depositary Governments.
3. This Agreement shall enter into force upon the deposit of instruments of ratification by five Governments including the Governments designated as Depositary Governments under this Agreement.
4. For States whose instruments of ratification or accession are deposited subsequent to the entry into force of this Agreement, it shall enter into force on the date of the deposit of their instruments of ratification or accession.

5. The Depositary Governments shall promptly inform all signatory and acceding States of the date of each signature, the date of deposit of each instrument of ratification of and accession to this Agreement, the date of its entry into force and other notices.
6. This Agreement shall be registered by the Depositary Governments pursuant to Article 102 of the Charter of the United Nations.

Article 8

Any State Party to the Agreement may propose amendments to this Agreement. Amendments shall enter into force for each State Party to the Agreement accepting the amendments upon their acceptance by a majority of the States Parties to the Agreement and thereafter for each remaining State Party to the Agreement on the date of acceptance by it.

Article 9

Any State Party to the Agreement may give notice of its withdrawal from the Agreement one year after its entry into force by written notification to the Depositary Governments. Such withdrawal shall take effect one year from the date of receipt of this notification.

Article 10

This Agreement, of which the English, Russian, French, Spanish and Chinese texts are equally authentic, shall be deposited in the archives of the Depositary Governments. Duly certified copies of this Agreement shall be transmitted by the Depositary Governments to the Governments of the signatory and acceding States.

In witness whereof the undersigned, duly authorized, have signed this Agreement.

Done in triplicate, at the cities of London, Moscow and Washington, the twenty-second day of April, one thousand nine hundred and

TREATY BANNING
NUCLEAR WEAPON TESTS
IN THE ATMOSPHERE
IN OUTER SPACE
AND UNDER WATER
(1963)

ENTERED INTO FORCE: 10 Oct 1963

The Governments of the United States of America, the United Kingdom of Great Britain and Northern Ireland, and the Union of Soviet Socialist Republics, hereinafter referred to as the "Original Parties",

Proclaiming as their principal aim the speediest possible achievement of an agreement on general and complete disarmament under strict international control in accordance with the objectives of the United Nations which would put an end to the armaments race and eliminate the incentive to the production and testing of all kinds of weapons, including nuclear weapons,

Seeking to achieve the discontinuance of all test explosions of nuclear weapons for all time, determined to continue negotiations to this end, and desiring to put an end to the contamination of man's environment by radioactive substances,

Have agreed as follows:

Article I

1. Each of the Parties to this Treaty undertakes to prohibit, to prevent, and not to carry out any nuclear weapon test explosion, or any other nuclear explosion, at any place under its jurisdiction or control:

   (a) in the atmosphere; beyond its limits, including outer space; or under water, including territorial waters or high seas; or

   (b) in any other environment if such explosion causes radioactive debris to be present outside the territorial limits of the State under whose jurisdiction or control such explosion is conducted. It is understood in this connection that the provisions of this subparagraph are without prejudice to the conclusion of a treaty resulting in the permanent banning of all nuclear test explosions, including all such explosions underground, the conclusion of which, as the Parties have stated in the Preamble to this Treaty, they seek to achieve.

2. Each of the Parties to this Treaty undertakes furthermore to refrain from causing, encouraging, or in any way participating in, the carrying out of any nuclear weapon test explosion, or any other nuclear explosion, anywhere which would take place in any of the environments described, or have the effect referred to, in paragraph 1 of this Article.

Article II

1. Any Party may propose amendments to this Treaty. The text of any proposed amendment shall be submitted to the Depositary Governments which shall circulate it to all Parties to this Treaty. Thereafter, if requested to do so by one-third or more of the Parties, the Depositary Governments shall convene a conference, to which they shall invite all the Parties, to consider such amendment.
2. Any amendment to this Treaty must be approved by a majority of the votes of all the Parties to this Treaty, including the votes of all of the Original Parties. The amendment shall enter into force for all Parties upon the deposit of instruments of ratification by a majority of all the Parties, including the instruments of ratification of all of the Original Parties.

Article III

1. This Treaty shall be open to all States for signature. Any State which does not sign this Treaty before its entry into force in accordance with paragraph 3 of this Article may accede to it at any time.
2. This Treaty shall be subject to ratification by signatory States. Instruments of ratification and instruments of accession shall be deposited with the Governments of the Original Parties–the United States of America, the United Kingdom of Great Britain and Northern Ireland, and the Union of Soviet Socialist Republics–which are hereby designated the Depositary Governments.
3. This Treaty shall enter into force after its ratification by all the Original Parties and the deposit of their instruments of ratification.
4. For States whose instruments of ratification or accession are deposited subsequent to the entry into force of this Treaty, it shall enter into force on the date of the deposit of their instruments of ratification or accession.
5. The Depositary Governments shall promptly inform all signatory and acceding States of the date of each signature, the date of deposit of each instrument of ratification of and accession to this Treaty, the date of its entry into force, and the date of receipt of any requests for conferences or other notices.
6. This Treaty shall be registered by the Depositary Governments pursuant to Article 102 of the Charter of the United Nations.

Article IV

This Treaty shall be of unlimited duration.

Each Party shall in exercising its national sovereignty have the right to withdraw from the Treaty if it decides that extraordinary events, related to the subject matter of this Treaty, have jeopardized the supreme interests of its country. It shall give notice of such withdrawal to all other Parties to the Treaty three months in advance.

Article V

This Treaty, of which the English and Russian texts are equally authentic, shall be deposited in the archives of the Depositary Governments. Duly certified copies of this Treaty shall be transmitted by the Depositary Governments to the Governments of the signatory and acceding States.

IN WITNESS WHEREOF the undersigned, duly authorized, have signed this Treaty.

DONE in triplicate at the city of Moscow the fifth day of August, one thousand nine hundred and sixty-three.

Index

MIX
Papier aus verantwortungsvollen Quellen
Paper from responsible sources
FSC® C105338

If you have any concerns about our products,
you can contact us on
ProductSafety@springernature.com

In case Publisher is established outside the EU,
the EU authorized representative is:
Springer Nature Customer Service Center GmbH
Europaplatz 3, 69115 Heidelberg, Germany

Printed by Libri Plureos GmbH
in Hamburg, Germany